NEXUS NETWORK JOURNAL   Architecture and Mathematics

Aims and Scope

Founded in 1999, the *Nexus Network Journal* (NNJ) is a peer-reviewed journal for researchers, professionals and students engaged in the study of the application of mathematical principles to architectural design. Its goal is to present the broadest possible consideration of all aspects of the relationships between architecture and mathematics, including landscape architecture and urban design.

Editorial Office

Editor-in-Chief
Kim Williams
Corso Regina Margherita, 72
10153 Turin (Torino), Italy
E-mail: kwb@kimwilliamsbooks.com

Contributing Editors
The Geometer's Angle
Rachel Fletcher
113 Division St.
Great Barrington, MA 01230, USA
E-mail: rfletch@bcn.net

Book Reviews
Sylvie Duvernoy
Via Benozzo Gozzoli, 26
50124 Firenze, Italy
E-mail: syld@kimwilliamsbooks.com

Corresponding Editors
Alessandra Capanna
Via della Bufalotta 67
00139 Roma Italy
E-mail: alessandra.capanna@uniroma1.it

Tomás García Salgado
Palacio de Versalles # 200
Col. Lomas Reforma, c.p. 11930
México D.F., Mexico
E-mail: tgsalgado@perspectivegeometry.com

Robert Kirkbride
studio 'patafisico
12 West 29 #2
New York, NY 10001, USA
E-mail: kirkbrir@newschool.edu

Andrew I-Kang Li
The Chinese University of Hong Kong
Shatin, N.T.
Hong Kong S.A.R. China
E-mail: andrewili@cuhk.edu.hk

Michael J. Ostwald
School of Architecture and Built Environment
Faculty of Engineering and Built Environment
University of Newcastle
New South Wales, Australia 2308
E-mail: michael.ostwald@newcastle.edu.au

Vera Spinadel
The Mathematics & Design Association
José M. Paz 1131 - Florida (1602), Buenos Aires, Argentina
E-mail: vspinade@fibertel.com.ar

Igor Verner
The Department of Education in Technology and Science
Technion - Israel Institute of Technology
Haifa 32000, Israel
E-mail: ttrigor@techunix.technion.ac.il

Stephen R. Wassell
Department of Mathematical Sciences
Sweet Briar College, Sweet Briar, Virginia 24595, USA
E-mail: wassell@sbc.edu

João Pedro Xavier
Faculdade de Arquitectura da Universidade do Porto
Rua do Gólgota 215, 4150-755 Porto, Portugal
E-mail: jpx@arq.up.pt

Instructions for Authors

Authorship

Submission of a manuscript implies:
- that the work described has not been published before;
- that it is not under consideration for publication elsewhere;
- that its publication has been approved by all coauthors, if any, as well as by the responsible authorities at the institute where the work has been carried out;
- that, if and when the manuscript is accepted for publication, the authors agree to automatically transfer the copyright to the publisher; and
- that the manuscript will not be published elsewhere in any language without the consent of the copyright holder.

Exceptions of the above have to be discussed before the manuscript is processed. The manuscript should be written in English.

Submission of the Manuscript

Material should be sent to Kim Williams
via e-mail to: kwb@kimwilliamsbooks.com
or via regular mail to: Kim Williams Books,
Corso Regina Margherita, 72,
10153 Turin (Torino), Italy

Please include a cover sheet with name of author(s), title or profession (if applicable), physical address, e-mail address, abstract, and key word list.

Contributions will be accepted for consideration to the following sections in the journal: research articles, didactics, viewpoints, book reviews, conference and exhibits reports.

Final PDF files

Authors receive a pdf file of their contribution in its final form. Orders for additional printed reprints must be placed with the Publisher when returning the corrected proofs. Delayed reprint orders are treated as special orders, for which charges are appreciably higher. Reprints are not to be sold.

Articles will be freely accessible on our online platform SpringerLink two years after the year of publication.

Nexus Network Journal

RECALLING EERO SAARINEN
1910-2010

Kim Williams and J. M. Rees, Editors

VOLUME 12, NUMBER 2
Summer 2010

KIM WILLIAMS BOOKS

Nexus Network Journal
Vol. 12
No. 2
Pp. 163-362
ISSN 1590-5896

CONTENTS

Eero Saarinen changed my life in a building. Since I am an architect I am susceptible to such things, and they are difficult to write about and conventionally excluded from our reasoned discourse about architecture. However; when Kim Williams announced this project – a collection of essays honoring Saarinen on the centennial of his birth – I jumped at the chance to assist. I cherish Eero Saarinen's work and hope to contribute, in some small way, to public conversation surrounding the built works.

Projects such as this special section of the *Nexus Network Journal for Architecture and Mathematics* call for an approach that is both bottom-up and top-down. Bottom-up, because one must always work within authors' particular interests; top-down, because the particular contributions must be integrated into a networked whole. The challenge is to craft a collection of given parts so that the assembly hangs together *and* reads separately. Surprisingly as this project developed, the top-down and bottom-up approaches morphed and then melded into the collection before you.

In top-down mode I was hoping for something relating architecture and music in Saarinen's work. It is a commonplace of architectural commentary to ally these arts since Matila Ghyka borrowed von Schilling's dictum "architecture is frozen music." As in most common notions, there is more than a grain of truth in the general observation, but general observations only go so far. Besides, there is a fluid quality to Saarinen's mature work that denies any kind of "petrified" form. Luckily David Foxe stepped forward with an essay entitled "Saarinen's Shell Game: Tensions, Structures, and Sounds at MIT", featuring a nuanced reading of Eero Saarinen's buildings at MIT. Based on the performance of music in two very different structures, Foxe's conclusion supports the complimentary roles of sound and structure which fosters both "external clarity and internal discovery."

Rachel Fletcher's contribution on Eero Saarinen's North Christian Church provides a different point-counterpoint. Juxtaposing quotes from Aline Saarinen's book of Eero's thoughts and drawings of the church, often superimposed with diagrams describing the architectural geometry, Fletcher develops the "merits of geometrical proportion." Most interesting however, is her conclusion that Saarinen responds "organically to the situation at hand."

Designing this issue on Eero Saarinen, I also hoped for an essay leading "where angels fear to tread." I am referring to the relationship between father and son: both designers, both visionaries, both strong personalities. Discussing this relationship, Susan Saarinen recounted a telling story. When Eliel Saarinen and Eero Saarinen decided to enter the competition for the Jefferson Memorial separately, the studio they shared was divided in half with a wall of blankets. Father and son actually used separate entrances for the duration of the competition phase of the project. The well known but often misinterpreted end of the story is also telling. The telegram notifying the winner was addressed to "E. Saarinen." It was the family that assumed it was addressed to Eliel.

These stories speak to competitive threads in the relationship between old-world father and new-world son. This is to be expected and indeed, who is surprised? But, it is not the whole story. Clearly the son defined himself in contradiction to the father, as sons are wont to do. Equally obvious is that there would have been no Eero

DOI 10.1007/s00004-010-0035-3; *published online* 27 May 2010

without Eliel. In this often told tale there are rivalries (the Jefferson Memorial competition, the General Motors site plan); there is the knowledge transfer between mentor and mentored (how to enter and win a competition, how to make "no small plans") and then there is the dialog between masterful equals. We will never be privy to the actual details of family dynamics yet we can conjecture some of this conversation in the built form of Christ Church Lutheran. This is exactly what Ozayr Saloojee does in his essay on "The Next Largest Thing: The Spatial Dimensions of Liturgy in Eliel and Eero Saarinen's Christ Church Lutheran, Minneapolis." His conclusion, in part, is "as a tribute to Eliel Saarinen with the education wing, Eero has managed to reflect his own attitude about architecture in a way that honors his father, but also his own vision." There is no higher compliment between two makers; father and son shine in the twice-reflected glow of a strangely symmetrical project.

The essay by Luisa and Victor Consiglieri, "Morphocontinuity in the work of Eero Saarinen," also speaks to strange symmetries: between form and structure, architecture and urban environment, inner and outer, visual perception and emotional movement. Their project, to give such notions a mathematical expression while retaining the messy complexity of agency, reminds me of the work of George David Birkhoff in *Aesthetic Measure*. Even though the effort is less than perfectly consonant (what discursive piece ever is?) the motivation – to join mathematics and biology and architecture – is uniquely suited to the *Nexus Network Journal*, which is after all a journal created to host just such difficult experiments.

The essay by Tyler Sprague, "Eero Saarinen, Eduardo Catalano and the Influence of Matthew Nowicki: A Challenge to Form and Function," is also devoted to a kind of experiment, this one about the way Matthew Nowicki speaks in the work of Eero Saarinen and Eduardo Catalano. As modern architecture developed, fresh vocabularies of form and of discourse emerged. Matthew Nowicki was in the vanguard of these developments. Nowicki supplied instances of each in the Livestock Pavilion (Raleigh N.C.) and in an essay titled "Origins and trends in Modern Architecture." Using these sources Sprague charts Nowicki's influence in works by Catalano and Saarinen arriving, appropriately enough, at conclusions regarding "how architectural discovery is often contextual."

Finally, the essay by Robert Osserman, "How the Gateway Arch Got its Shape," is an insightful romp through the entwined histories of science, mathematics and architecture. From a top-down editorial point of view "The geometry of the Gateway Arch" was exactly the essay I had already decided not to accept. As a native of the state in which the arch stands, I had read too many comments like "its shape is a (fill in the blank) catenary." In fact, much of the excitement in this project for me was in rediscovering other masterworks by Eero Saarinen: John Deere Headquarters, Milwaukee War Memorial. I thought I knew all about the arch, having effectively grown up with it. Boy was I wrong. We hold in special regard those whose works expose us to our own unfounded prejudice. This is what Professor Osserman's essay did for me – not only for the geometry of the arch but also for Galileo (he really did understand the difference between a parabolas and catenaries) and Robert Hooke (who was not closed-minded in the arrogant sort of way like I had always assumed). Beyond simply correcting my prejudicial ignorance Bob Osserman's essay couples a word picture of the arch's form – flattened catenary – with a mathematical formalism, effectively joining an intuitive understanding with a technical description in a scrupulous presentation that takes care to call reader's attention to open questions.

This essay is altogether satisfying and I can only hope that it, along with all the other works in this section devoted to Eero Saarinen on the centennial of his birth, find wide readership.

At the end of this process I want to thank all the essayists for their hard work and express my gratitude to Kim Williams for picking up the ball. Regarding Saarinen, I am again reminded of my debt to him. I can honestly say that he has changed my life twice: once when I was a student of architecture, through the TWA terminal, and then again as a more mature practitioner reviewing a life in design well lived, cut tragically short. Two things Eero Saarinen repeatedly demonstrated will always inspire: the courage to approach each project as its own functional and aesthetic problem – its own opportunity, style be damned, and that space is a rooted surface at play in light. Thank you E.S. We owe you much.

J.m.Vees

About the guest editor

J. M. Rees, colorist, architect, writer, is editor of The Sixth Surface: Steven Holl Lights the Nelson-Atkins Museum of Art (2007) and Urban Stories of Place (2006). He practices architecture in Kansas City, Missouri. Exhibitions include Conceptual Play at the Greenlease Gallery (KCMO), the virtual nomad at the Linda Hall Library for the History of Science and Technology (KCMO), and Manhattan Miniature Golf at P.S.1 (Queens, NY). He enjoys collaboration and believes that works of imagination are always more effective when they respond to multiple conditions with layers of signification.

1618 Summit St.
Kansas City, Missouri 64108 USA
abcjmr@swbell.net

* * *

The celebration of the centennial of Eero Saarinen's birth 1910-2010 provides a welcome opportunity to examine the work of this great twentieth-century master builder. Other research included in this issue takes us to different parts of the world and different epochs, and demonstrates a link between the ancient and modern attempts to unite architecture and mathematics.

Indian professor R. Balasubramaniam, who sadly passed away before he could see this issue in print, studied the two significant funerary gardens of the Mughal period. In "On the Modular Design of Mughal Riverfront Funerary Gardens", he proves that the modular designs of these Mughal funerary gardens were based on *Arthasastra* units, and sets forth a novel mathematical canon to analyze the dimensions of Mughal architecture.

In "Discontinuous Double-shell Domes through Islamic eras in the Middle East and Central Asia: History, Morphology, Typologies, Geometry, and Construction," Maryam Ashkan and Yahaya Ahmad present a developed geometric approach for deriving the typologies and geometries of discontinuous double-shell domes in Islamic architecture. Common geometric attributes are created using a corpus of twenty one domes that were built in the Middle East and Central Asia, beginning from the early through to the late Islamic periods.

Giulio Magli takes us to Peru during the age of the Incas. In "At the Other End of the Sun's Path: A New Interpretation of Machu Picchu," he challenges the standard interpreation of Machu Picchu by means of a critical reappraisal of existing sources and a re-analysis of existing evidence, leading to a new interpretation of this mysterious city.

Tessa Morrison examines "The Body, the Temple and the Newtonian Man Conundrum", a fascinating look at a little-known aspect of Isaac Newton: his interest in architecture, and especially, his knowledge of Vitruvius.

This issue is rounded out with a book review and a conference report. Areli Marina reviews Nigel Hiscock's *The Symbol at Your Door: Number and Geometry in Religious Architecture of the Greek and Latin Middle Ages*. Eir Grytli reports on a conference held in Trondheim, Norway, in honor of Professor Emeritus Staale Sinding-Larsen's eightieth birthday.

Finally, I am very proud to announce that the Nexus Network Journal has been accepted into Thomson's ISI database for scholarly journals, and will be listed in the Art & Humanities Citation Index from 2009 forward. The Arts & Humanities Citation Index is a multidisciplinary index to the journal literature of the arts and humanities. It fully covers 1,160 of the world's leading arts and humanities journals, and also indexes individually selected, relevant items from over 6,800 major science and social science journals. The NNJ will receive an impact factor rating starting in 2011. A special thank you to all those who have contributed to the NNJ, from its founding in 1999 to the present, for establishing the high standard of scholarship that characterizes the journal. I am very grateful to you all.

Whether your own interests are historic or modern, you'll find fascinating reading in this Summer 2010 issue of the Nexus Network Journal.

Kim Williams

Robert Osserman

Robert Osserman
Mathematical Sciences Research
Institute
17 Gauss Way
Berkeley, CA 94720 USA
ro@msri.org

Keywords: Eero Saarinen, Gateway
Arch, catenary, parabola, weighted
catenary, Robert Hooke, Galileo
Galilei, history of mechanics

Research

How the Gateway Arch Got its Shape

Abstract. Robert Osserman examines Eero Saarinen's Gateway Arch in St. Louis in order to shed light on what its exact shape is, why it is that shape, and whether the various decisions made during its design were based on aesthetic or structural considerations. Research included discussions with engineers and architects who worked with Saarinen on the project. The paper concludes by noting some questions that are still unanswered.

Introduction

Fig. 1. Eero Saarinen's Gateway Arch, St. Louis. Photo courtesy of Historic American Engineering Record, Library of Congress, Prints and Photograph Division

Much has been written on the subject of Eero Saarinen's most widely known creation and architectural landmark, the Gateway Arch in St. Louis. Nevertheless, it is difficult to find complete and authoritative answers to some of the most basic questions:

1. What is the shape of the Arch?
2. Why is it the shape that it is?
3. Of the various decisions that had to made, which were based on esthetic, and which on structural considerations?

These questions raise issues of a purely mathematical nature that are of independent interest. A number of them have been treated elsewhere [Osserman 2010] and will be referred to as needed. Here we focus on the basic questions that are our central concern.

The Shape of the Arch

The terms most often used in describing the shape of the Gateway Arch are *parabola, catenary,* and *weighted catenary.* We shall discuss each of those terms in some detail later on, but let us note that they all represent mathematical idealizations of physical phenomena. A parabola, as Galileo demonstrated, is the shape of a path traversed by a projectile subjected only to its initial impetus together with the force of gravity, in the absence of air resistance. A catenary is the shape assumed by a hanging chain or a flexible cord of uniform density. A weighted catenary is the shape assumed by a hanging chain whose links vary in size or weight, or by a flexible cord of variable width, or variable density material.[1]

An equally basic feature of the shape of the Arch is what we shall call, borrowing a term from computer, television, and movie screens, its "aspect ratio": that is, the ratio of width to height of the inside dimensions of the smallest picture frame that can hold the full frontal view of the Arch; or, in architectural drawing terms, a front elevation of the Arch. Aspect ratios are usually expressed in a form such as "3-to-2" or "3:2."

An occasional source of confusion stems from the fact that both the parabola and the catenary extend to infinity. In both cases there is, up to scaling, a single curve, but arcs of a single parabola or catenary can look very different from each other, depending on the part that one chooses, and the corresponding aspect ratio of the given arc. In fact, in both cases, one may choose arcs with any aspect ratio one wishes. Fig. 2 illustrates how to obtain a variety of aspect ratios from a single catenary.

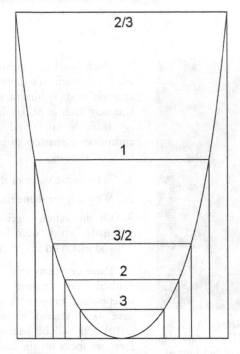

Fig. 2. Aspect ratios for a catenary

A further potential source of confusion is that Saarinen changed the aspect ratio of his Arch design between the time that he originally won the competition, and the final design used in the construction. The original design was for an arch that was 630 feet wide and 590 feet high, whose aspect ratio was therefore approximately 21:20, or just shy of square. By the time the arch was built, Saarinen had kept the original width, but raised the height to 630 feet, exactly matching its width. According to architect Bruce Detmers,[2] who worked with Saarinen on the Arch, it was typical of Saarinen to favor simple geometric figures, such as a square, and therefore not surprising that he would modify his design to obtain the 1:1 aspect ratio. According to other sources, the principal reason that Saarinen increased the height of the arch was that between the time that he won the original competition in 1948, and the beginning of construction in the early 1960s, new and higher buildings had been added to the surrounding landscape, which forced Saarinen to modify his own plans so that the Arch would clearly dominate its surroundings. It is undoubtedly the case that both of these factors played a role in his final choice of the exact 1:1 aspect ratio.

It may be worth noting that the 1:1 aspect ratio is a very old tradition in architecture. In medieval times it was known as *ad quadratum*. The original 1386 specifications for the cathedral of Milan is an example (cf. [Heyman 1999: 19]).

The contrast between the shape of a parabola and a catenary is clear if we choose segments of each with a 1:1 aspect ratio (fig. 3).

In fig. 3, the parabola is the pointier inner curve and the catenary the rounder outer curve.

Fig. 3. Parabola and Catenary with aspect ratio 1:1

History of the problem

Commemorative arches designed to be lasting monuments date back thousands of years, with a number still standing from the days of ancient Rome. However, it does not appear to have been until 1675 that the question was raised – at least in print – of what shape an arch *should be,* from a mathematical and structural point of view. At that time, one of the leading scientists of the day,[3] Robert Hooke, provided his answer to the question in the form of an anagram of the Latin phrase: *ut pendet continuum flexile, sic stabit contiguum rigidum inversum,* or, "as hangs the flexible line, so but inverted will stand the rigid arch." Hooke was famous for his ability to devise physical demonstrations and experiments illustrating both old and newly discovered scientific principles, and was specifically commissioned to do that for the recently created Royal Society. In the case of the ideal shape of an arch, he realized that what one wants is that the slope of the arch at each point exactly match the combined horizontal and vertical forces acting on that part of the arch – the vertical component being due to gravity from the weight of the portion of the arch lying above the point, and the horizontal force being simply transmitted unchanged along the arch. At the apex of the arch, there is no weight above it, hence no vertical component, and the force is simply the horizontal one of the two sides of the arch leaning against each other. As a result, the slope at that point is zero. As one moves down along the arch, the vertical force keeps increasing, and the slope of the arch must increase accordingly.

One simple consequence of this general reasoning is that at the base of the arch, the horizontal force continues to be present – a fact well known to builders who created flying buttresses and other devices to counter that force – and as a consequence, the bottom of the arch should never be strictly vertical, but rather, angled outwards. As it happens, probably not by chance, Hooke recorded his dictum at the time when both he and Christopher Wren were among the chief architects in charge of surveying the damage, and of rebuilding London after the disastrous fire of 1666. Hooke shared his insight with Wren, who immediately applied it to his design of St. Paul's cathedral, which features an interior dome that is the first to be angled out at its base, rather than vertical (cf. [Heyman 1999: 40-41]).

But the main thrust of Hooke's observation was that the exact shape of an ideal arch could be obtained by simply hanging a chain (or "flexible line") and recording the form that it takes. The equilibrium position would be determined by a balance of vertical and horizontal forces that are the mirror image of what they would be in the case of the arch, with the role of gravity reversed: the vertical component at each point is determined by the weight of the chain below it that needs to be supported, while the horizontal component is simply transmitted unchanged along the chain.

Left unanswered by Hooke is the obvious mathematical question raised by his purely physical solution to the ideal shape of an arch: "what is the equation of the curve that is formed by a hanging chain?" Wherever this subject is mentioned in print, one is likely to find a statement to the effect that Galileo was the first to raise the question, and that he answered incorrectly that the shape that the chain would take was a parabola. Both of those statements merit a closer look.

Galileo's discussion of the shape of a hanging chain appears in print in his last, and in the opinion of many, his best, book, *Two New Sciences*. The book takes the form of four days of dialogue on a wide array of subjects. The first day introduces the first "new science": the analysis of the strength of beams and columns from a scientific and mathematical point of view. It is often considered as the incubator of the modern fields of structural engineering and strength of materials. During the second day, Galileo has his chief protagonist, Salviati, propose two practical methods of drawing a parabola [Galilei 1974: 143]. The first is by rolling a ball along a tilted metal mirror, and the second is by hanging a fine chain, and marking points along it. In other words, he does not pose the question of the shape taken by a chain, he simply assumes that it takes the form of a parabola, based on both theoretical reasoning, using the decomposition of the forces acting on it in the vertical and horizontal directions that led him to deduce the parabolic path of a projectile, as well as on experimental evidence: drawing a parabola on a vertical board, and observing the shape of a hanging chain that seems to follow the same curve.

Few commentators note that Galileo returns to the same question on the fourth day of his dialog, and explicitly states that the similarity is only approximate. He also explains the analogy between the forces acting on a projectile, and those on a hanging chain. Here is the exact quotation:

> Salviati: The curvature of the line of the horizontal projectile seems to derive from two forces, of which one ... drives it horizontally, while the other ... draws it straight down. In drawing the rope, there is [likewise] the force of that which pulls it horizontally, and also that of the weight of the rope itself, which naturally inclines it downward. So these two kinds of events are very similar.
>
> ...
>
> But I wish to cause you wonder and delight together by telling you that the cord thus hung, whether much or little stretched, bends in a line that is very close to parabolic. The similarity is so great that if you draw a parabolic line in a vertical plane ... and then hang a little chain from the extremities ... , you will see by slackening the little chain now more and now less, that it curves and adapts itself to the parabola; and the agreement will be the closer, the less curved and the more extended the parabola drawn shall be. In parabolas described with an elevation of less than 45°, the chain will go almost exactly along the parabola [Galilei 1974: 256-7].

In short, Galileo never poses the question of precisely what shape is taken by the hanging chain, and he contents himself with noting that it provides a close approximation to a parabola, especially when the parabolic arc that one draws is near to the vertex where the shape is relatively flat. For example, a 45° parabola would be represented by the equation

$$y = \tfrac{1}{2} x^2, \quad -1 \leq x \leq 1,$$

which fits in a rectangle of width 2, and height ½, so that its aspect ratio is 4:1. Fig. 4 shows the parabola together with a catenary having the same aspect ratio.

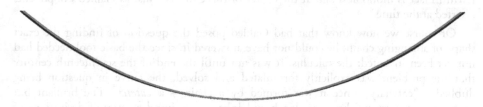

Fig. 4. 45° parabola and catenary with the same aspect ratio: 4:1

It is often pointed out that unlike the case of a freely hanging cable which will take the form of a catenary, when a cable supports a roadway that is much heavier than itself and whose weight is distributed evenly along the horizontal rather than equally along the cable, the cable will take the shape of a parabola. That will be the case for the cables joining the two towers of a suspension bridge.

Fig. 5. The Golden Gate Bridge

In the case of the Golden Gate Bridge, shown in fig. 5, the central cables form curves that make an angle of almost exactly 30° with the horizontal at the points where they leave the towers. However, as fig. 6 makes clear (or as one can show by a direct calculation), the parabola and approximating catenary are virtually indistinguishable in this case.

Fig. 6. 30° parabola and corresponding catenary

The fact that the overwhelming number of references to Galileo's treatment of this subject declare him to have been in error, and ignore the fact that he ends up getting it exactly right, would seem to reflect an extra degree of pleasure in catching him out in any perceived slip-up. In fact, the first page to come up on a Google search of "Galileo wrong" starts with an item from NASA entitled "Was Galileo wrong?" and ends with a NOVA television program entitled "Galileo's battle for the heavens: his big mistake." The first of these entries questions whether the acceleration of gravity is the same for all varieties of materials – a fundamental tenet of science ever since Galileo, and certainly true within the limits of measurements available to him (and many subsequent generations). The last has to do with an actual mistake: his theory that tides were caused by the rotation of the earth. In between these two items are a series of references to a book entitled "Galileo was Wrong: The Church was Right," devoted to proving that the Earth in fact *is* motionless and at the center of the Universe, just as Galileo's opponents asserted at the time.[4]

Of course we now know that had Galileo posed the question of finding the exact shape of a hanging chain, he could not have answered it, since the basic tool needed had not yet been invented: the calculus. It was not until the end of the seventeenth century that the problem was explicitly formulated and solved, the curve in question being dubbed a "catenary" since it was formed by a chain – a *catena*.[5] The brilliant but cantankerous Bernoulli brothers, Jacob and Johann, produced a series of derivations of the equation for the catenary along with a series of barbs at each other's reasoning, such as Jacob's parody of an argument of Johann that he describes as somewhat like proving that a pebble is stone by reasoning that "every man is stone; every pebble is a man; therefore every pebble is stone."[6] Whatever the shortcomings of the reasoning, they did indeed arrive at the correct conclusion. The equation for the catenary can be written in the form we now call the hyperbolic cosine,

$$y = \cosh x = \tfrac{1}{2}\,(e^x + e^{-x})$$

in suitable choice of coordinates.[7]

The design of the Arch

The Gateway Arch was only one component – although clearly the most dramatic one – of Eero Saarinen's winning entry in the 1947 open competition for the design of a "Jefferson National Expansion Memorial," dedicated to Thomas Jefferson – to his vision of an America stretching across the continent, and his Louisiana Purchase, which roughly doubled the size of America at the time. By a truly strange coincidence, the first use of "catenary" recorded in the Oxford English Dictionary is by the future President Thomas Jefferson, in a letter dated December 23, 1788 to Thomas Paine, recommending the use of a catenary arch rather than a circular one for a 400-foot span iron bridge that Paine is proposing to build. Jefferson would certainly have been inordinately pleased, as an architect himself, to know that the principle of the catenary would form the basis of what was to be the largest monument in America in the twentieth century, in honor of Jefferson himself.

The Jefferson Memorial competition attracted virtually all the major architects in America, including Eero's father Eliel, a much more famous architect at the time than his 37-year old son. At least one part of the confusion involving the shape of the Gateway Arch stems from the fact that the competition was held in two parts, first narrowing down to the top five contenders, and then – after significant revisions were made – choosing the winner among those five. In addition, there was a gap of 15 years between the time that Saarinen's proposal was chosen and when construction actually began, during which time still further changes were adopted. And even after construction had started, alterations were made, as one can see from different sets of blueprints in the Saarinen archives at Yale University. Further confusion arises from a certain degree of carelessness in terminology that can be seen, for example, in a dictionary of architecture where a picture of an arch is accompanied by a caption referring to it as a "catenary or parabolic arch" [*Visual Dictionary of Architecture* 2008: 41], not aware of, or not interested in, fine distinctions. Whatever the reason, newspapers and architectural journals that reported or commented on the shape of Saarinen's winning design for the Gateway Arch, referred to it without exception as a parabola.[8] By chance, among the letters that Saarinen received was one from H. E. Grant, head of the Department of Engineering Drawing at Washington University in St. Louis, asking for more details about the "parabolic arch," since Grant wanted to use it as an example in a book on descriptive geometry that he was writing at the time. In his response, dated March 24, 1948, Saarinen says:

> The arch actually is not a true parabola, nor is it a catenary curve. We worked at first with the mathematical shapes, but finally adjusted it according to the eye. I suspect, however, that a catenary curve with links of the chain graded at the same proportion as the arch thins out would come very close to the lines upon which we settled."[9]

In other documents Saarinen left considerable room for ambiguity. In 1959 he wrote:

> The arch is not a true parabola, as is often stated. Instead it is a catenary curve – the curve of a hanging chain – a curve in which the forces of thrust are continuously kept within the center of the legs of the arch.[10]

Whether by "catenary curve" he means an actual catenary, or is trying to hedge a bit, is not clear. In an unpublished transcript dating from 1958-59, he describes the Arch as "an absolutely pure shape where the compression line goes right through the center line of the structure directly to the ground. In other words, a perfect catenary" [Eero Saarinen 2006: 343].

If one consults the Wikipedia entry for "catenary" one finds – at least at the time of this writing (June 2009):

> The Gateway Arch in Saint Louis ... follows the form of an inverted catenary. ... The exact formula
> $$y = -127.7 \text{ ft} \cdot \cosh (x/127.7 \text{ ft}) + 757.7 \text{ ft}$$
> is displayed inside the arch.

The equation given is indeed that of a catenary, and it has the desired dimensions of 630 feet in both height and width, but it is not the equation that was used in the construction of the arch, and that appears on all the blueprints and specifications.[11]

At the Arch itself, a sheet provided by the National Park Service, under whose jurisdiction the monument falls, is headed "Equation for the catenary curve of the centroid of arch cross-section." Below that is the same equation that appears on the blueprints. We shall return to the equation shortly, but it is definitely not a catenary.

In 1983, an article entitled "Is It a Catenary? New questions about the shape of Saarinen's St. Louis Arch," appeared in an architectural journal [Crosbie 1983]. The author, Michael J. Crosbie, explained that despite widespread belief that the Gateway Arch is shaped like a catenary, it is not. Instead, it is a "weighted catenary," the form taken by a "weighted chain" in which the various links are of different weight, rather than all the same. In mathematical terms, it amounts to using a "flexible line" with a density that is variable rather than constant. It appears to be uniformly true that any description of the shape of the Gateway Arch that tries to be accurate will use the term "weighted catenary" or something similar.

We are now in a position to state precisely the mathematical questions that arise.

1. How much information is conveyed by the expression "weighted catenary"? More precisely, what range of curves may be obtained by suitable choice of a variable density?

2. What was the basis for the choice of the particular "weighted catenary" chosen by Saarinen? Was it primarily on esthetic or on structural/mathematical grounds?

3. The arch is designed so that its cross-sections are equilateral triangles. Again the question whether that choice was esthetic or structural.

4. The size of the cross-sections varies according to a precise formula, with the sides of the equilateral triangles growing from 17 feet at the apex to 54 feet at ground level. What determined those dimensions, and what is the significance of the specific formula used for the tapering? And once again, were those choices motivated by esthetic or mathematical considerations?

The answer to the first question can be formulated quite simply: the term "weighted catenary" conveys essentially no useful information; virtually any curve that one can picture as a possible candidate[12] for a weighted catenary can actually be reproduced by a suitably weighted chain.[13] Among those are, for example, a parabola and a circular arc, as well as the catenary-like curve of the actual arch. In all of those cases one can explicitly say how to distribute the weight so that the chain, or flexible line, will hang in exactly the desired shape. In other words, saying that the shape of the Gateway Arch is a weighted catenary is technically correct, but essentially devoid of content. Furthermore, it is not descriptive, in that it says nothing about the actual shape of the Arch, but only about the method used to produce it.

On the other hand, there is a very simple description of the precise shape of the Arch. It is what we call a "flattened catenary" and consists simply of a catenary that has been shrunk uniformly in the vertical direction by a given amount. It will have a vertical axis of symmetry, just as the catenary does, and if we choose that axis to be the y-axis, then the equation of a flattened catenary takes the form

$$y = A \cosh Bx + C, \tag{1}$$

where $A, B > 0$. Note that if we take the catenary curve $y = \cosh x$, and scale it uniformly up or down in size, we get the equation $By = \cosh Bx$, while flattening it by a uniform compression in the vertical direction would give the equation $y = D \cosh x$, where

$$0 < D < 1.$$

In other words, equation (1) represents a catenary if and only if $A = 1/B$. The constant C corresponds simply to a translation in the vertical direction, and we if choose coordinates so that the vertex of the curve is at the origin, then $C = -A$. Putting all this together, we see that by setting $D = AB$, equation (1) takes the form

$$y = D(1/B)(\cosh Bx - 1) \tag{2}$$

so that equation (2) represents a catenary with vertex at the origin that has been flattened vertically by the factor D.

The centroid curve of the Gateway Arch is of exactly this form, where A and B are specified numerical constants:

$$A = 68.7672, \quad B = .0100333, \tag{3}$$

so that the flattening factor D is given by

$$D = .69 \tag{4}$$

In other words, the catenary is shrunk in the vertical direction by just under a third. The effect is to "round it out" somewhat more at the vertex. The y-coordinate in equation (2) for the Gateway Arch represents the vertical distance *down* from the vertex of the centroid curve. The numerical values given in (3) correspond to distances measured in feet.

It is worth noting that shrinking a curve in the vertical direction is exactly equivalent, up to uniform scaling, to expanding it uniformly in the horizontal direction – precisely what is often done to reformat a film done with one aspect ratio to fit it onto a wider screen with a different aspect ratio (fig. 7).

It is also important to observe that since the flattening degree D for the Arch given by (4) is a little over 2/3, and since the centroid curve of the Arch is intended to have an aspect ratio of approximately 1:1, ("approximate" because it is the outer silhouette of the Arch that is designed to have aspect ratio 1:1, and the same will not be true of the centroid curve, as we discuss more fully below,) we must start with a portion of the catenary having an aspect ratio of approximately 2:3 and either shrink it vertically by about 2/3, or stretch it horizontally by about 3/2 (fig. 8).

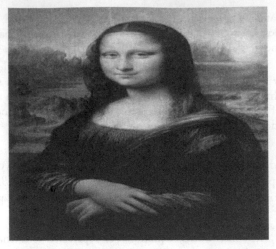

Fig. 7. Wide-screen Mona Lisa

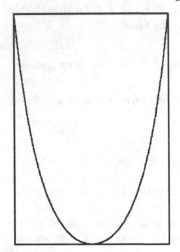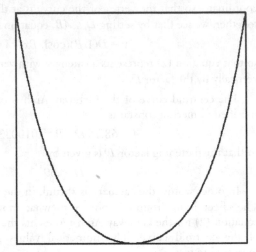

Fig. 8. Re-formatting to obtain a 1:1 aspect ratio

It is ironic that Saarinen has been attacked from both sides, with some commentators criticizing the shape of the arch for being too purely mathematical to be taken seriously as a work of art, while others take him to task for not being "pure" enough. Michael Crosbie [1983] notes the fact that "the curve is not a perfect catenary," adding "— a pure form that would seem more attractive to Saarinen...". Hélène Lipstadt is more severe: "The purity is illusory and the explanation deceptive (if not deceitful), since he knowingly had his engineers calculate an impure, weighted catenary" (quoted in [Merkel 2005: 244, note 36]).

Clearly, a catenary is no higher on the scale of purity than, say, a parabola or a circular arc, which are "weighted catenaries." The equation for the parabola is perhaps the simplest of all equations for a curve, consisting as it does of a polynomial with a single (quadratic) term. A circular arc is one step more complicated, involving square roots, and hence, algebraic functions. The catenary, built out of exponential functions, which are not even algebraic, but so-called transcendental functions, would seem to lie relatively far afield in the domain of mathematical purity. Presumably what is meant by the criticism is that the equation Saarinen used as a basis for his arch, on the suggestion of Hannskarl Bandel, one of his engineers, suffers from being *almost*, but not quite, a catenary, involving the insertion of rather arbitrary-looking numerical constants.

To criticize Saarinen for not using a perfect catenary, however, is to miss the point of Hooke's original insight. The catenary is the shape one obtains using a uniform chain or "line" and is therefore the shape one wants for an arch of uniform thickness. For a relatively small arch, uniform thickness would not be unreasonable, but for the kind of monumental structure proposed by Saarinen for his Arch, it would be almost unthinkable, both on esthetic and structural grounds. One would naturally want the cross-section of the arch to be the least near the apex, where it has little or nothing to support and where any additional weight adds to the strength requirements of everything below. Conversely, the parts of the arch closest to the ground have the most weight to support, and would naturally have the largest cross-sectional size. As a consequence, the correct shape would not be the mirror image of a uniform chain, but of a correspondingly weighted chain, as Saarinen points out in the quotation given above. That principle was clearly understood early on. A famous illustration from 1748 in a report by Giovanni Poleni on the cracks in the dome of St. Peter's shows a cross-section of the dome, together with the shape of a hanging chain that was weighted proportionally to the loads of the corresponding sections. Poleni used the result to deduce that the structure was basically sound, despite the cracks that had appeared [Heyman 1999: 38-39].

In short, a "pure catenary" is the ideal shape for an arch of uniform thickness, and a flattened catenary is the ideal shape for an arch that is tapered in a certain precise (and elementary) manner. In the case of a weighted chain, we may express the amount of weighting by a density function $\rho(s)$, where the weight of any arc of the chain is the integral of $\rho(s)$ with respect to arclength s over that arc. One can show (see [Osserman 2010] that a given flattened catenary of the form $y = f(x) = A \cosh Bx + C$ is obtained from a density function ρ which, when expressed in terms of x, takes the form

$$\rho(s(x)) = (ay + b)/(ds/dx) = (a\,f(x) + b)/\sqrt{(1 + (f'(x))^2)},$$

where the coefficients a and b are determined by the coefficients A, B, and C of $f(x)$, up to an arbitrary multiple (since multiplying the density by an arbitrary positive constant just changes the total weight, but not the shape of a hanging chain.) What this means physically is that the weight of any portion of the chain lying over a segment of the x-axis, given by $\int\rho(s)ds$, is equal to $\int(ay+b)dx$ over the given x-interval.

That brings us to the answer to our fourth question: how the shape of the Arch — which is to say, the shape of its centroid curve — is related to the degree of tapering.

As we have noted, the centroid curve of the Gateway Arch is a flattened catenary whose equation is of the form (1), where the x-axis is horizontal, the origin is at the vertex, or highest point of the curve, and y represents the distance down from the vertex. We denote by h the height of the curve, and by w its width. The fact that $y=0$ when $x=0$ translates to the condition $C = -A$, so that the equation takes the form

$$y = A\,(\cosh Bx - 1). \tag{5}$$

Then the fact that $y = h$ when $x = \pm w/2$ implies

$$h = A(\cosh Bw/2 - 1). \tag{6}$$

If we introduce the notation

$$R = \cosh Bw/2 \, ,$$

then equation (6) becomes

$$A = h/(R-1) \tag{7}$$

while from the definition of R,

$$B = (2/w) \cosh^{-1} R \, . \tag{8}$$

The coefficients A, B that enter into the equation for the centroid curve given on the blueprints and all the official documents for the Gateway Arch have exactly this form (fig. 9).

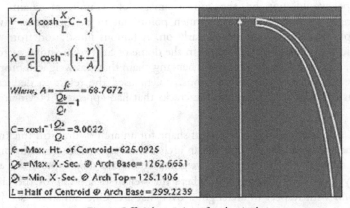

Fig. 9. Official equations for the Arch.
Source: http://www.nps.gov/jeff/planyourvisit/mathematical-equation.htm

We see that their somewhat strange appearance arises simply from the condition that the curve pass through the points $(0,0)$ and $(w/2, h)$. The only question is the value of the constant R. By its definition, we have $R > 1$, and conversely, if we choose an arbitrary number $R > 1$, then equation (5) with coefficients defined by (7) and (8) represents a flattened catenary passing through the given points. We find further that the degree of flattening D of the curve is given by

$$D = AB = (h/w)(2 \cosh^{-1}R)/(R-1) \tag{9}$$

Or, equivalently,

$$D = (1/r)(2 \cosh^{-1}R)/(R-1) \tag{10}$$

where we have denoted by r the aspect ratio of the centroid curve,

$$r = w/h \, . \tag{11}$$

In other words, the flattening coefficient D is completely determined by the desired aspect ratio r and the – so far – arbitrary constant R.

We are finally in a position to spell out precisely the relation between the shape of the Arch and the basic geometric decisions that were made. In particular we see that the complicated-looking values of the coefficients A and B in the equation of the Arch are direct consequences of a few simple choices. Those choices are:

(i) the centroid curve be a flattened catenary;

(ii) the orthogonal sections be equilateral triangles with one vertex pointing inward;

(iii) the triangles at the top and the bottom be 17 and 54 feet on a side respectively;

(iv) the overall height and width each be 630 feet;

(v) the constant R in equations (7) and (8) be given by $R = Q_b / Q_v$, where Q_b is the cross-sectional area of the triangle at the base, and Q_v the corresponding area at the top.

From (ii) and (iii) it follows that the cross-sectional areas at the vertex and the base are

$$Q_v = 125.1406, \quad Q_b = 1262.6651, \qquad (12)$$

so that according to (v), the value of R is a little over 10. Also from (ii) and (iii) we can calculate the height h and width w of the centroid curve needed to make the outside dimensions of the arch satisfy (iv), and they in turn determine the aspect ratio r. Inserting all these quantities into the equations (7), (8), and (9), gives the numerical values of the coefficients A and B that we noted earlier, as well as the degree D of flattening of the curve.

The one remaining choice that must be made is a formula for the tapering from top to bottom of the Arch. The decision there was to make the cross-sectional area $Q(x,y)$ at the point (x,y) of the centroid curve a linear function of y, interpolating between the given values (12) at the top and bottom. In other words,

(vi) $Q(x,y) = Q_v + (Q_b - Q_v) y/h$.

Conditions (i) – (vi) together provide a complete mathematical description of the basic design of the Gateway Arch. Its actual construction, with an outside skin consisting of flat stainless steel plates, requires dividing this basic shape into sections,[14] each bounded by a pair of equilateral triangles, and replacing the curved surface defined by the equations by three flat pieces.[15] The result is also completely defined mathematically, but is an artifact of construction, intended to so closely approximate the curved surface that one is not even aware, for example, that the inner curve of the Arch, its intrados, is not a smooth curve, but a polygonal one. That is indeed the case, and we limit our discussion here to the curved surface of the Arch that underlies the construction.

What remains to answer is whether the choice of the shape itself was dictated largely by structural or by esthetic considerations. When this question was put to Bruce Detmers, who worked with Saarinen on the Arch, his answer was unequivocal: Saarinen's choice was purely esthetic. According to Detmers, Saarinen worked with a large variety of shapes, many of them made on the elaborated Hooke principle, by hanging a weighted chain, or a piece of rubber cut out with a varying width to gain the same effect. The resulting shapes were then used, upside down, to make sizeable models of the arch, and Saarinen would examine the models from all angles. He ended up rejecting them all, as either too flat or too pointed, or otherwise unappealing. He then turned to the engineer Hannskarl Bandel for further ideas, and it was Bandel who suggested the shape that Saarinen found to his liking and that was the one actually used for the design of the Arch. (As a parenthetical note, the key role of the engineers throughout the process, from design to completion, is all-too-often forgotten in accounts of the Arch and its construction.)

The question that arises, if one accepts that Saarinen chose the shape of the Arch on purely esthetic grounds, is "What role, if any, was played by mathematics?"

Bruce Detmers notes two such roles. First, mathematics was critical as a means of communication. It is one thing to draw a shape and to make models that are 10 or 12 feet high, but if one arrives at a pleasing design, then it is still necessary to convert that design into specific instructions to provide those who must manufacture and assemble the parts. There is no obvious way to make the transition from a sketch or a model that must be scaled up by several orders of magnitude and described with absolute precision in order to arrive at the full-sized structure. But as soon as the shape is defined by a mathematical equation, one may build it on any scale that the available materials and methods will allow.

Second, Detmers also alluded to Saarinen's predilection for perfect geometric shapes like a square or a circle. In the case of the Arch, as we have already noted, he decided that he would like its proportions to be such that it could be exactly inscribed in a square.

An additional, and crucial role of mathematics was to calculate the ideal shape of the fully three-dimensional structure whose central curve had the form decided on. That comes down to a process that is the reverse of starting with a hanging chain and adopting the form it takes in order to build an arch. Here one starts with a desirable shape, and has to determine the degree of tapering that would be structurally ideal to go with that shape. But as we have noted, to virtually any shape defined by a mathematical equation, one can assign a weighting that will produce exactly that shape.

The history of the particular equation suggested by Bandel for the Arch goes back to the mid-nineteenth century. It arose in trying to determine the ideal shape for an arch that supports an earth bridge – a bridge formed by simply piling up dirt above the supporting arch to make a flat road surface. The steepness of the arch at each point should be determined by the balance of the horizontal and vertical forces at the point; the horizontal force is just transmitted along the arch and is unchanged from point to point, while the vertical force results from the weight of the column of dirt directly above the point, and is therefore proportional to the height of the road above that point on the arch. From that one can show that the ideal shape must be a flattened catenary (see, for example [Heyman 1982: 48-49]).

In the case of the Gateway Arch, it does not support any other structure. However, the effect of the tapering is that the weight being supported by each cross-section increases steadily as one moves from the top to the bottom of the arch, just as it does for an earth bridge. What emerges then is a simple answer to our second and fourth questions regarding the shape of the Arch and the amount of tapering: the two are inextricably entwined, as are the esthetic and structural decisions that have to be made.

The reality is that far more comes into play than these basic considerations. In any project of the magnitude of the Gateway Arch, economic and political factors are inevitably going to be involved. One striking example is the appearance of the George

Washington Bridge, with its two silvery towers of interlaced steel girders. According to the original design, those towers were to be covered in stone in the manner of the bridge's famous forerunner, the Brooklyn Bridge, and the steel framework one now sees were to be the invisible support. However, the money for the project ran out, and generations of New Yorkers, along with millions of out-of-town visitors, have engraved in their minds a very different-looking bridge than the one envisaged by its designers. The Eads Bridge across the Mississippi that now forms a backdrop to the Gateway Arch was as dramatic for its time as the George Washington Bridge across the Hudson several generations later. In both cases, the length of the bridges and the size of the river spanned broke new ground, and in both cases the bridges linked two adjacent states. In the case of the Eads Bridge, connecting Missouri to Illinois, powerful political factors entered in, affecting the final design of the bridge.[16]

Even on the structural level, many factors must be considered beyond the ideal shape for supporting the weight of the Arch itself. There are many forces besides gravity that must be taken into account, including those arising from potential gale-force winds, from earthquakes, and from uneven expansion and contraction as the sun heats up certain parts more than others. In addition, one must assure that the Arch is not susceptible to vibrations at certain resonant frequencies that might lead to structural failure, as happened in the famous case of the Tacoma Narrows Bridge.

Finally, there are many additional decisions that must be made, where esthetic and structural decisions are intermingled. For the overall effect, the most important was undoubtedly the choice of stainless steel for the outer surface, reflecting blue skies and clouds during the day, shining brilliantly against a dark sky when lit up at night. And right down to important details, such as the choice of whether to "brush" the steel horizontally or vertically, and what method to use for attaching adjoining steel plates.

That leaves us with one remaining question to be answered – question 3 about the choice of a triangular cross-section. The answer forms part of the subject of our final section.

Some controversial aspects of the Gateway Arch

No project as monumental as the construction of the Gateway Arch is apt to come to completion without arousing serious controversy. At the least, there are always matters of taste and a ready chorus of critics whose job it is to criticize. In the particular case of the Arch, there were four flags raised against Saarinen that are worthy of note.

The first was raised at the very outset, during the original competition: "What good is it?" A great deal of (public) money is to be spent on a structure with no "purpose" other than merely symbolic. (That is a criticism familiar to most mathematicians who engage in and teach "pure" mathematics, often with no practical applications in sight.)

It should be said that the Gateway Arch was designed with the express purpose of putting St. Louis back on the map after a long period of decline from its nineteenth-century glory as the second largest port in America and the "Gateway to the West." In that capacity, it has been wildly successful, attracting millions of visitors each year, and joining the Washington Monument and the Eiffel tower as among the most widely known symbols of their kind.

The unqualified success of the Arch, both symbolic and practical, as a draw of both attention and visitors to St. Louis, has pretty well answered this objection. Beyond that, many are struck by the beauty of the shape, and the issue then turns on the value of a piece of art or architecture that is admired and gives pleasure of a purely esthetic kind.

The second controversy arose during the construction of the Arch, when serious worries were expressed about its potential to fail and to collapse in what would have been a major catastrophe. That conflict and its resolution have been beautifully described by one of the principal figures involved, George Hartzog.[17]

The third regards our question about the choice of a triangular cross-section. Saarinen's original design for the Arch had a quadrilateral cross-section. Between the first and second phase of the competition, one of his colleagues, the sculptor Carl Milles, suggested that a triangular cross-section would be more attractive. Saarinen liked the idea and adopted it in his final designs, but he failed to publicly credit Milles with the suggestion, a failure that caused many hard feelings and much anger.[18]

The fourth controversy, and most interesting from many points of view, arose shortly after the proposal submitted by Saarinen and his team of co-workers was awarded the commission to build the monument. An article in the *New York Herald Tribune* displayed side-by-side pictures of the design submitted by Saarinen and a poster for an International Exposition that was planned for 1942 in Rome. The headline was "St. Louis Arch For Memorial Called Fascist." In fact, the Rome exposition was explicitly billed as commemorating the twentieth anniversary of Fascism in Italy, and Mussolini's "March on Rome" in 1922. The poster features an arch designed by the Italian architect Adalberto Libera that appears to be on much the same scale as that envisaged by Saarinen, and seems to resemble it strongly in shape. Saarinen vehemently denied knowing anything about the earlier proposed arch, which in fact was never built, due at least in part to the advent of World War II. He further noted that a "simple form, based on the natural laws of mathematics," could no more be co-opted by a political movement than the Washington Monument could be faulted for the fact that its design is that of ancient Egyptian obelisks that were built by slaves.

Ironically, Saarinen's defense – that the form of his Arch was a standard one that could well have been used by anybody – has the effect of diminishing the originality of his design. On the other hand, it provides fairly convincing evidence that he knew nothing about the earlier design. The resemblance on the poster turns out to be purely an artifact of the particular angle at which the Libera Arch is depicted. The actual plans submitted by Libera show that his proposal was for a strictly semi-circular arch, with no resemblance to Saarinen's, other than the overall similar scale, and of course the fact that both arrived at the same idea of a monumental arch as a symbolic gateway – in Libera's case, to the proposed Exposition and the associated newly rebuilt area of Rome, still known as EUR: Esposizione Universale Roma – and for Saarinen, to the broad Western expanse of America across the Mississippi River.

An additional irony is that when Saarinen's design acquired renown after he won the competition, Libera accused him of plagiarism, whereas Libera had himself been similarly charged, because another team of Italian architects had suggested much the same design for an entry arch to the EUR somewhat earlier than Libera.

The final irony is that even if World War II had not intervened, it is not clear that Libera's arch would have been built, or that if built, it would have been able to resist collapse. In fact, a semi-circle is far from an ideal shape for a monumental arch. Serious doubts were expressed at the time. Some of those doubts concerned the choice of materials and structural details of Libera's proposal. Others are intimately connected with Robert Hooke's original insight connecting the shape of an arch with that of a hanging chain. As we noted earlier, one consequence, put to use immediately by Christopher Wren, is that one wants the base of the arch to hit the ground at some angle, and not vertically, as is the case of a semi-circle. The larger the size of the arch, the more vital such considerations become. That is true for an arch of uniform thickness and the corresponding chain of uniform weight. But it is equally true for a weighted chain of any sort.

We have also noted earlier that although it may seem completely counter-intuitive, it is possible to weight a chain in such a manner that it will hang in the form of a perfect circular arc, provided only that it is less than a full semi-circle. That weighting will in turn indicate the ideal tapering of the arch for maximum structural stability. However, Libera's original specifications for his proposed arch give no indication that he was aware of the mathematics needed to determine the correct weighting or degree of tapering.[19]

Concluding remarks

The paintings and the book by Jasper Johns devoted to catenaries [2005] make it clear that the shape of a hanging chain evokes an esthetic response quite apart from any attempt to analyze that shape by mathematical means.

In the case of architecture, the need to combine esthetic values with mathematical and structural principles is one that dates back to antiquity, and is certainly apparent in the surviving Greek temples. In some cases, mathematics and esthetics were viewed as interchangeable. As is apparent from our discussion here, in the case of the Gateway Arch they are, if not equivalent, then at least inextricable. It is clear from his letters and his various explanations, that Saarinen was aware at a very early stage that the structural requirement for maximal stability translated into keeping the line of thrust aligned as nearly as possible with the slope of the Arch, which in turn determined the relevant shape. Having begun with that principle, and tried out the resultant shape, he turned it around, modified the pure catenary shape "according to the eye," and then left it to his (superb) engineering team to apply the mathematical principles that would produce the desired shape and satisfy the structural needs.

Viewed more broadly, the role of the architect is one that has provoked a surprisingly wide range of responses. At one extreme, there is the assessment reported in the book *Brunelleschi's Dome* by Ross King [2000: 157-158], that both ancient and medieval authors "assigned architecture a low place in human achievement, regarding it as an occupation unfit for an educated man." He cites both Cicero and Seneca in support of that view. At the other extreme, we have the view that "it is hardly surprising that, for the ancients, the image of the architect has demiurgic connotations" [Benvenuto 1991: Introduction, xix]. The author, Edoardo Benvenuto, goes on to say:

> In a famous dialogue of Paul Valéry[20] this "nearly divine" aspect is expressed in the following words of Phaedro to Socrates when speaking of his friend the architect Eupalinos:[21] "How marvelous, when he spoke to the workmen! There was no trace of his difficult nightly meditation. He just gave them orders and numbers.

(To which Socrates responds, "God does just that".)

A twentieth-century version of this story is reported by Bruce Detmers, whose duties as a young architect working with Saarinen on the Gateway Arch included long hours spent translating the engineers' formulas for Saarinen's Arch into long columns of numbers generated, in the absence of a modern computer, on an old Marchand calculator borrowed from the accounting department. These numbers were then passed on to those responsible for manufacturing the various components needed in the construction of the Arch.

Detmers recounts one incident that may resonate as a distant echo of Phaedro and Eupalinos. He was working away at two or three o'clock in the morning, when Saarinen wandered in, and asked what he was doing. When Detmers explained, Saarinen's laconic response was, "Keep going."

Additional remarks and open questions

We elaborate here on some of the points noted briefly in our earlier paper [2010], and conclude with some remaining questions.

1. It is often noted that viewers of the Gateway Arch tend very strongly to see it as being taller than wide, despite the fact that its height and width are both exactly equal at 630 feet. This effect is generally described as an optical illusion. But is it, in fact, an illusion?

While it is true that the *outside* curve, or extrados, is the same width as height, one's eye is as likely to fix on the inside curve, or intrados, or on something in between. In those cases, the curve one sees, or extrapolates, is in fact considerably taller than wide. The reason is simply that the arch is much thinner at the top than at the base, and that even further, the base width gets subtracted off twice between the outer and inner curves. In precise numerical terms, the inner curve is 615.3 feet high and 536.1 feet wide, so that the height is 15% greater than the width. It is likely the case that the impression of the Arch being taller than wide is partly due to this fact and partly a true optical illusion.[22]

2. It is clear that Saarinen was aware of Hooke's dictum relating the shape of a standing arch to that of a hanging chain, and that the goal was to build an arch where the direction of the line of thrust at each stage of the arch is as near as possible to the slope of the arch itself. He was further aware that the principle applied whether the chain was uniform or variable, forming a "weighted chain." In choosing the shape of a flattened catenary, rather than a pure catenary, he was able to satisfy two distinct desires simultaneously – first, to arrive at a shape more pleasing esthetically to him, since is was more rounded at the top, and second, to have the corresponding arch satisfy the structural needs of greater thickness and strength toward the bottom, while thinner and more slender near the top. If one compares both the pure and the flattened catenaries with a parabola, one sees that the parabola is even pointier at the top than the catenary (fig. 10).

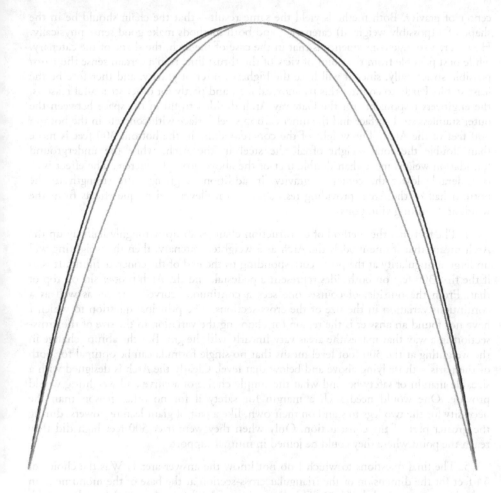

Fig. 10. Parabola, Catenary, and Flattened Catenary with Aspect Ratio 1:1

It is fortunate (or is it something deeper?) that Saarinen found the parabola less pleasing from a purely esthetic viewpoint, since following Hooke's dictum in that case would require greatest weight at the vertex, and therefore an arch that was thickest at the *top*, and slimming down toward the bottom—hardly what one would want from a structural point of view. In fact, it is clear that if one starts with a uniform chain that hangs in the shape of the catenary, then to make it pointier at the vertex one would want to increase the weight in the middle, while to make it rounder or flatter near the vertex, one would increase the weight near the ends.

Fig. 10 shows all three curves together, each with aspect ratio 1:1. The parabola is the innermost, pointiest curve, and the catenary the intermediate one, while the outer curve is a flattened catenary with the same degree D of flattening as that used in the Arch.

3. There are at least two ways to arrive mathematically at the shape of a hanging chain. The first is the one we have been invoking, in which the object is to keep the thrust line always in the direction of the chain. The other is by using the method known as the calculus of variations to determine the shape of the chain that will give the lowest

center of gravity. Both methods yield the same result – that the chain should be in the shape of a (possibly weighted) catenary – and both methods make good sense physically. However, a strange consequence is that in the case of the arch, the shape of the catenary, while best possible from the point of view of the thrust line, is in a certain sense the *worst* possible structurally, since it will have the highest center of gravity, and therefore be the least stable. Partly to counter this paradoxical fact, and partly for other structural reasons, the engineers responsible for the Gateway Arch decided to fill in the space between the outer stainless steel surface and the inner carbon steel surface with concrete in the bottom 300 feet of the Arch. The weight of the concrete alone in the bottom 300 feet is more than double the total weight of all the steel in the Arch, while the underground foundation weighs more than double that of the above-ground concrete. The effect is to considerably lower the center of gravity, in addition to giving extra strength to the bottom half of the Arch, providing resistance to cantilever and torque effects from the wind, and limiting vibrations.

4. The fact that the method of construction changes abruptly roughly halfway up the Arch means that if one models the Arch as a weighted catenary, then the weighting will undergo a singularity at the point corresponding to the end of the concrete filling. It is as if the first 300 feet on both sides represent a pedestal, and the Arch proper sits on top of that. From the outside, of course, one sees a continuous curved surface, as well as a continuous variation in the size of the cross-sections. The principal question to which I have not found an answer is the reason for choosing the variation in the size of the cross-section in a way that makes the areas vary linearly with height. But the abrupt change in the weighting at the 300-foot level means that no single formula can be optimal for both of the parts – those lying above and below that level. Clearly the Arch is designed with a sizeable margin of safety beyond what the simple choice of a curve and weighting would provide. One would need such a margin for safety if for no other reason than the necessity for the two legs to stand on their own, like a pair of giant leaning towers, during the greater part of the construction. Only when they were over 500 feet high did they reach the point where they could be joined in mutual support.

5. The final questions to which I do not know the answer are: 1) Was the choice of 54 feet for the dimension of the triangular cross-section at the base of the monument an esthetic or a structural decision? (The dimension of 17 feet at the top, I have been told, was simply the smallest felt to be possible, allowing for the construction of an observation platform where visitors can stand up and admire the view from the top through the windows built into the Arch.); 2) What is the basis of the choice of the ratio Q_b / Q_v for the parameter R that enters into the basic equations (7) – (10)? I hope to return to these questions at a later occasion.

Acknowledgments

The most interesting and accurate mathematical discussions of the Gateway Arch that I am familiar with are those by William V. Thayer [1984], and a chapter (in Portuguese) of a volume of computational activities and projects designed to highlight applications of the calculus [Figueiredo et al. 2005]. I am particularly indebted to Bill Thayer for providing a wealth of documents as well as a number of significant leads in my investigation of this problem; also to Charles Redfield, one of the main engineers who worked on the Arch and who provided me with a number of relevant documents, Bruce Detmers, an architect who worked with Saarinen on the Arch, John Ochsendorf for enlightening discussions of his paper [Block et al. 2006] on arches and related matters,

Jacques Heyman for prompt responses to a number of questions arising from his many books and articles about arches, Hélène Lipstadt and Jack Rees who offered a number of further valuable leads, and Robert Moore of the National Park Service, historian of the Gateway Arch. I also thank Jennifer Clark, archivist at the Old Court House in St. Louis, and Laura Tatum at the Saarinen Archive at Yale University

Notes

1. One can also treat a cord from which are hung a series of literal weights, in which case the cord will take a kind of polygonal shape. However, we shall not consider that case here, since it is not too relevant to the curve of the Gateway Arch.
2. This and all further references to statements by Bruce Detmers come from conversations with him in March 2008.
3. Robert Hooke had the great misfortune to be a contemporary of Isaac Newton, and on two counts. First of all, Newton's brilliance and monumental achievements simply overshadowed any and all potential rivals among scientists. But perhaps even more so, because Newton developed one of his notorious lifetime grudges against Hooke, and did everything possible – which was quite a bit – to denigrate and belittle Hooke and his achievements.
4. See also [Huerta 2006]. The issue here is that although Galileo is precisely correct in his great insight that one cannot scale up any structure an arbitrarily large amount, since the weight goes up as the cube of the scaling factor and the cross-sectional area of the supports – say columns for a building, legs of an animal, or trunk of a tree – increases by only the square, nevertheless he is "wrong" not to note that on the scale of normal buildings, strength is not the key issue, but stability, as in Hooke's dictum, and in that case what counts is the geometrical shape, invariant under scaling.
5. Strictly speaking, the term used was the Latin word, *catenaria*, and the English word "catenary" did not occur until somewhat later; this will be discussed shortly. An excellent and detailed history of the "catenary problem" is given by Clifford Truesdell in The Rational Mechanics of Flexible or Elastic Bodies, 1638-1788; Introduction to *Leonhardi Euleri Opera Omnia*, Vol. X et XI Seriei Secundae, Lausanne, Orell Füssli Turici 1960; see also further references in footnote 7 below. (Surprisingly, Truesdell also fails to cite Galileo's correct description of the catenary as an approximation to a parabola.)
6. "*Tout homme est pierre; tout caillou est homme; donc tout caillou est pierre.*" From a letter of August 11, 1697; Jacob Bernoulli, *Opera*, pp. 829-839; reprinted in *Die Streitschriften von Jacob und Johann Bernoulli*. Birkhäuser Verlag 1991, pp. 356-364 (see p. 361). See also pp. 1-114 of this volume for an overview in English of the contents, and pp. 117-122 for a list of "The Polemic Writings of Jacob and Johann Bernoulli on the Calculus of Variations" that are reproduced in the book.
7. Other names that should be mentioned in this brief history are Philippe de la Hire and David Gregory. De la Hire's *Traité de méchanique* from 1695 addresses the question of how to distribute weights along a chain to attain "a figure curved the way you wish it to be" (proposition 123), and he makes explicit the connection between arches and variably weighted chains. David Gregory discovered independently the relation between weighted chains and arches and described them in a letter later published in the *Philosophical Transactions* in August 1697. For more on both of these major contributions, see [Benvenuto 1991: 321-329]. Another good historical reference is [Bukowski 2008] (the usual misleading statements about Galileo are repeated here, but it is instructive to see how early they originated and were propagated by successive generations).
8. At least that is the case for the dozens of articles, as well as congratulatory telegrams that are on file in the Saarinen Archives at YaleUniversity.
9. Copies of Grant's letter and Saarinen's response are in the Saarinen archive. It is the only place I have found where Saarinen describes his winning design with as much specificity.
10. Quoted in [Crosbie 1983].
11. Whether this formula (or, for that matter, any other formula) is displayed inside the Arch as claimed, is something I have not been able to determine. It is not impossible, since the precise

shape of the Arch was modified so many times during the planning process, but it is unlikely, because the decision not to use an exact catenary was made very early – long before construction began.

12. The precise conditions for the graph of a curve to be representable as a weighted catenary are that it be convex, and that it makes a non-zero angle with the vertical at its endpoints. See [Osserman 2010].

13. Johann Bernoulli was already aware of the connection between arches and a chain of variable weight in 1698; see, for example, [Benvenuto 1991: Pt. II, 327].

14. There are 71 such sections on each leg of the arch, according to the specifications.

15. This is further complicated by a note on the blueprints that for the top 41 sections, the "exterior skin to be curved into a continuous smooth surface." This corresponds to roughly the upper half of the Arch, where the curvature is the greatest. It is also above the 300-foot level where there is another transition in the mode of construction, as we discuss more fully later on; below that level the stainless steel outer skin is backed by a layer of concrete.

16. Both of these examples are described in fascinating detail in the book *The Tower and the Bridge* [Billington 1985].

17. George B. Hartzog, Jr. was Director of the National Park Service at the time. See the chapter on the Gateway Arch in [Hartzog 1988], especially pp. 52-56.

18. See, for example, [Coir 2006]: "Milles strongly advised Saarinen to drop his initial plans to fashion the arch with a quadrilateral section in favor of a more sculpturally pleasing triangular section. Eero's failure publicly to recognize Milles's contribution to the monument angered the Swedish sculptor, who thereafter severed contact with Saarinen" [2006: p. 41 and footnote 40, p. 43].

19. On the other hand, some of Libera's sketches for the arch show a radical thickening near ground level, which would seem to indicate at least an intuitive understanding of the need for a suitable weighting to support the shape of a circular arc. From an esthetic point of view that thickening may be seen as producing a more dramatic effect for Libera's arch than for Saarinen's, or else as far less graceful. (Perhaps both.)

20. Benvenuto is referring to Valéry's "Eupalinos ou l'architecte" [1960: 83].

21. Eupalinos was the engineer to whom is ascribed (by Herodotus) the construction of the famous 3400 foot long tunnel on Samos, excavated from both ends, meeting in the middle.

22. A number of experiments seem to confirm that we tend to consistently overestimate lengths in the vertical direction over equal ones that are horizontal. See for example [Wolfe et al. 2005: 967-979] and further references given there.

Bibliography

BENVENUTO, Edoardo. 1991. *An Introduction to the History of Structural Mechanics. Part II: Vaulted Structures and Elastic Systems.* New York, Springer-Verlag.

BILLINGTON, David. 1985. *The Tower and the Bridge.* Princeton: Princeton University Press.

BLOCK Philippe, Matt DEJONG, John OCHSENDORF. 2006. As Hangs the Flexible Line: Equilibrium of Masonry Arches. *Nexus Network Journal* 8, 2: 13-24.

BUKOWSKI, John. 2008. Christiaan Huygens and the Problem of the Hanging Chain. *College Mathematics Journal* 39, 1: 2-11.

COIR, Mark. 2006. The Cranbrook Factor. Pp. 29-43 in Eero Saarinen, *Shaping the Future.* New Haven: Yale University Press.

CROSBIE, Michael J. 1983. Is It a Catenary? New questions about the shape of Saarinen's St. Louis Arch. *AIA Journal* (June 1983): 78-79.

Eero Saarinen: Shaping the Future. 2006. Eeva-Liisa Pelkonen and Donald Albrecht, eds. New Haven: Yale University Press.

FIGUEIREDO, Vera L.X., Margarida P. MELLO, Sandra A. SANTOS. 2005. *Cálculo com Aplicações: Atividades Computacionais e Projetos.* Coleção IMECC, Textos Didáticos.

GALILEI, Galileo. 1974. *Two New Sciences.* Stillman Drake, trans. Madison: University of Wisconsin Press.

HARTZOG, George B., Jr. 1988. *Battling for the National Parks.* Kingston (Rhode Island): Moyer Bell.

HEYMAN, Jacques. 1999. *The Science of Structural Engineering*. London: Imperial College Press.
HEYMAN, Jacques. 1982. *The Masonry Arch*. Chichester: Ellis Horward Limited.
HUERTA, Santiago. 2006. Galileo was Wrong! The Geometrical Design of Masonry Arches. *Nexus Network Journal* 8, 2: 25-51.
JOHNS, Jasper. 2005. *Catenary*. New York: Steidel Publishing.
KING, Ross. 2000. *Brunelleschi's Dome: How a Renaissance Genius Reinvented Architecture*. Penguin Books.
MERKEL, Jayne. 2005. *Eero Saarinen*. London: Phaidon Press.
OSSERMAN, Robert. 2010. Mathematics of the Gateway Arch. *Notices of the American Mathematical Society* 57, 2: 220-229.
THAYER William V. 1984. *The St. Louis Arch Problem*. Module 638 of UMAP: Modules in Undergraduate Mathematics and its Applications.
VALÉRY, Paul. 1960. Eupalinos ou l'architecte. In *Oeuvres*, Vol. 2, Paris: Bibliotèque de la Pléiade/Éditions Gallimard.
The Visual Dictionary of Architecture. 2008. Lausanne: Ava Publishing.
WOLFE, Uta, Laurence T. MALONEY, and Mimi TAM. 2005. Distortion of perceived length in the frontoparallel plane: Tests of perspective theories. *Perception and Psychophysics* 67, 6: 967-979.

About the author

Robert Osserman is professor emeritus of mathematics at Stanford University and Special Projects Director at the Mathematical Sciences Research Institute in Berkeley. His research interests within mathematics have centered on geometric questions in a variety of fields, including complex function theory, partial differential equations, Riemann surfaces, minimal surfaces, isoperimetric inequalities, and ergodic theory. He has also written about interactions between mathematics and areas of the arts and humanities, including music, visual arts, theatre, and film, as well as a book on geometry and cosmology, *Poetry of the Universe: a Mathematical Exploration of the Cosmos* (Anchor Books, 1996).

David M Foxe

125 Warren Street
Newton Centre, MA 02459,
USA
dmfoxe@yahoo.com

Research

Saarinen's Shell Game: Tensions, Structures, and Sounds at MIT

Keywords: Eero Saarinen, Kresge
Auditorium, MIT Chapel,
music, acoustics, religious
buildings, auditorium,
shell/membrane design,
structural design, graphic statics,
interdisciplinary, student life,
Massachusetts Institute of
Technology

Abstract. For all of the criticism of Eero Saarinen's Kresge Auditorium and MIT Chapel, they exist as a highly focused moment of deliberate experimentation with geometric form in materials old and new which both contrasted the typical forms of rational modernism and resonated deeply with the modernist quest for the incorporation of novel structures. This paper explores the metaphorical and literal tensions through three dichotomies: geometrical ones with implications for acoustics, programmatic ones with implications for use, and structural ones between the appearances and actual structural actions of the architecture. I seek to illuminate how the geometric issues of both buildings relate to structural optimization. I also approach the Auditorium and Chapel from the roles of having been a performer and composer of instrumental and vocal music for both spaces while earning degrees at MIT in architecture. The simple act of listening defies one's typical expectations in both spaces, and the dichotomies of geometry, use, and structure illuminate the relationship of sound to place in these architectural spaces.

Prelude: Sites and Sounds

Nestled in a small stand of London Plane trees planted west of Massachusetts Avenue stands the MIT Chapel designed by Eero Saarinen. Designed in tandem with the grassy oval and the spherical Kresge Auditorium, Saarinen's chapel creates a small, closed, cylindrical form in the midst of its domed and rectilinear neighbors on the MIT campus. The exterior form (modified in recent decades by causeway-like ramps moderating the level change at the entry) encloses a space that is purposefully hidden. Across Kresge Oval, the auditorium seeks a less contemplative and more forthright appearance from Massachusetts Avenue; its curtain wall broadcasting views of the lobby since its entry is at the grade of the broadly sloping turf.

While the chapel is relatively recent, merely a half-century old, it hearkens back to more primal architecture concepts, focusing on simple elements of water and light. While sited within viewing distance of a major river, Saarinen's design includes a shallow pool of water that functions as a reflecting pool to the exterior as well as reflecting light to the interior. While the exterior remains a quiet, closed cylinder surrounded by still water, the interior of the chapel is a ring of undulating brick ripples.. Provided it has not been seasonally drained from the concrete "moat," the water reflects light through small horizontal curved windows up onto the brick; these ever-changing patterns of light are the primary contact the interior space has with the outside world, the light having been mediated by the water's surface. For many traditions, the single overhead oculus piercing the curved, black, night-like ceiling is an appropriate connection to divine connotations (fig. 1).

Nexus Network Journal 12 (2010) 191–211   NEXUS NETWORK JOURNAL – VOL.12, No. 2, 2010 **191**
DOI 10.1007/s00004-010-0027-3; *published online* 4 May 2010
© 2010 Kim Williams Books, Turin

Fig. 1. MIT Chapel, main aisle (view toward Bertoia screen). Photograph by the author

Fig. 2. MIT Chapel, exterior approach. Photograph by the author

While the water reflects light upward, the oculus admits a stream of light downward onto the altar and circular marble slabs, bringing the "light of the world" to a surrounding environment of darkness.[1] While the quality of light from the oculus is diffused by the grate-like structure therein, the altar screen is a powerful connection from the light source through the space to the ground. Descending from the heavens to the realm of humanity, the deceptively simple yet intricate structure uses wire and metal plates to reflect light (see fig. 7). Designed by Harry Bertoia (another alumnus of the Cranbrook School in Bloomfield Hills, MI, which Eliel Saarinen designed and his son Eero attended), this structure uses the individuality of the rough metal plates to be beautiful not only in surface detail, but also as a whole when perceived from further away in the space. Since it is supported at its top and bottom by the building, it does not tend to read as "sculpture" separate from the architecture, but rather as an emphatic focal point within the aesthetic whole. The sculpture mediates the transition from the light of above to the illuminated marble below, reflecting the mediation between God and man at times of worship.

The use of light and water for elegant architectural principles is not limited to the round interior of the chapel space. In contrast to the open sunlight of Kresge Oval, the trees around the chapel provide a filtered mosaic of light and shadow on the brick cylinder as well as the ground (fig. 2). The water surrounding the chapel creates a physical and visual separation penetrated only by the bridge-like passage of the entryway. Rather than having the spatial feel of a traditional narthex, this low, translucent corridor provides a transitional, intermediate space that sets the viewer on a linear course down the chapel's main axis to the altar and oculus (fig. 3).

Fig. 3. Chapel, entry lobby. Photograph by the author

The chapel also benefits from Saarinen's careful treatment of exterior space. The aforementioned axis is aligned, not with the center of the auditorium, but with the off-center placement of the south auditorium entrance (fig. 4). Furthermore, the whole chapel-auditorium combination has been placed to the south of the main axis of MIT.

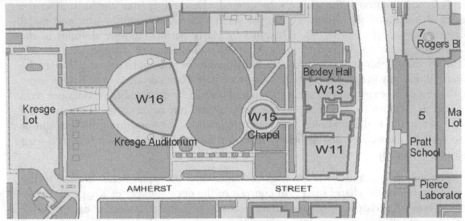

Fig. 4. Detail of a map of the MIT campus, showing the placement of Kresge Auditorium and the MIT chapel

By shifting the axial emphasis, a bold and unusual concept for compositions of circular forms, Saarinen creates asymmetrical outdoor spaces amongst the trees, giving each building its own sense of place and identity without competing for prominence. Even the use of brick in the chapel and the exterior terrace of the auditorium helps to build a visual relation to the curved brick forms of Aalto's Baker House visible beyond (see fig. 10). Prior to the existence of Kresge Auditorium, the MIT Chapel, and the raised oval of landscaped space, this area behind the apartment buildings (now Bexley Hall) on Massachusetts Avenue was a flat, open space.[2] While the chapel may now appear more "central" to the overall campus, it predates all of the subsequent dormitories on Amherst Street as well as the student center. Therefore, it was slightly more on the outskirts of campus at the time of construction, but foreshadowed the growth of West Campus as a residential area. Along with the formal development of the humanities as a school and of students' living opportunities on campus, Saarinen's buildings constituted a concurrent attempt to develop the opportunities for spiritual space in an urban, secular environment.

In the highly crafted exterior space between the chapel and auditorium, the surrounding forms are mainly horizontal, except for the single element atop the chapel. Theodor Roszak's design for the Bell Tower cannot be appreciated at the same proximity or level of detail as the interior sculpture due to its physical location, but it contributes a bold upward gesture that relieves the staid, earthbound forms of the chapel and auditorium.[3] Although one MIT description calls it a symbol of "the history and authority of three major religious persuasions," [MIT, *Art and Architecture* 1988: 45] this seems to contradict and limit the more broad definition of an inherently 'multidenominational' or 'nondenominational' chapel, and is not supported by Saarinen's own statements. The bell tower can also be perceived in campus lore as an abstracted 'rocket to the heavens,' or a gestural steeple, but its essence seems to exist not in its overt symbolism, but rather in its verticality as the one element that aspires above and beyond the surrounding tree canopy, extending the worshipful interior space into the exterior environment and the atmosphere above.

The auditorium has no such pinnacle; it is often recognized and published for its iconic copper-paneled roof, but it did not initially have such a covering; at the time of its opening the stark whiteness of the membrane prompted critical reception.[4] Even though the initial membrane failed and reroofing was necessary less than a decade after initial construction,[5] the copper has persisted in a role as the metallic analogue of brick in having a surface which becomes richer with age.

> Kresge Auditorium, MIT's performance and rehearsal hall, is the chapel's fraternal – and far from identical – twin. Kresge's graceful roof, sheathed in copper, its triangular plan and its glass-and-steel windows contain a hive of activity. A little theater, a concert hall and rehearsal rooms within are used for everything from drama to dance to music performances, as well as symposia and science, technology and engineering conferences [Wright 2005].

Moreover, these program spaces designed for sound are column-free: the main hall seating over 1200 and the theatre for 200, and the remainder of subterranean spaces for rehearsal, storage, and service purposes are supported from below. Early site schemes foresaw the demolition of Bexley Hall along Massachusetts Avenue and a more formal relationship to both the main entrance at 77 Massachusetts Avenue at street level or via other overpasses and underpasses, none of which were built. Saarinen's vision of a consistent northern edge was compromised by the looming brutalist symmetries of Eduardo Catalano's Student Center in 1968, but have been improved by the elegant glazed scrim of the Zesiger Fitness Center completed by Roche/Dinkeloo (the successor firm to Saarinen's) in 2002 (fig. 5). The character of the auditorium is less baroque and fine-grained in the scale of its detailing than the chapel, its curtain wall mullions less delicate and its glazing more conventional, but even the exterior components of its support structure at the corners define the lobby's interior geometries within the enclosure.

Fig. 5. Kresge Auditorium, exterior view, approach from Massachusetts Avenue (Zesiger Center at right). Photograph by the author

1 Geometries: Formal and Informal

First, the techniques of form introduce geometrical complexities into the aforementioned first-level simplicity. Both the chapel and the auditorium are constructed on a circular plan and a solid of revolution: a cylinder and a sphere. The solid is shown plainly in the case of the chapel, while the sphere (112-ft radius) is truncated into an eighth, i.e., one rotated quadrant of three-dimensional space such that its center (origin) is far underground (fig. 6). The chapel is embellished by surrounding the cylinder with its moat, while the auditorium ripples with further repeated concentric brick plinths, stepping down to the lower land elevation to the west.

Yet the circles are presented as both a *solution* and as a *problem* to be solved: Saarinen speaks of the cylindrical plan as a result of site explorations to create a form that maximizes its similarity as seen from many vantage points, and as a function of minimizing the maximum interior distance [Saarinen 1962: 34-6]:

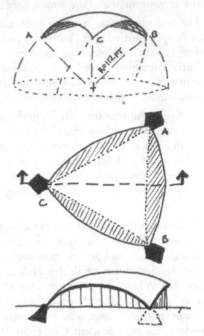

Fig 6. Kresge Auditorium, sketch of geometry
(sketch by the author,
after [Saarinen 1962: 127])

> We believed that what was required was a contrasting silhouette, a form which started from the ground and went up carrying the eye around its sweeping shape. Thus, a domed structure seemed right. There were other reasons, too, that influenced us toward a dome. There was the large dome of Welles Bosworth's central building at MIT. A dome is an economical way of covering an area with concrete. It is a shape which provides a pleasant atmosphere for an auditorium. And a thin-shell concrete structure seemed an appropriate form to express the spirit of this advanced school of technology. [...]

> After many experiments, exploring different shapes in the site plan, the round cylindrical form seemed right. The circular shape also seemed right in plan – for this was basically a chapel where the individual could come and pray and he would be in intimate contact with the altar. [...]

> The interior wall was curved, both for acoustical reasons, and to give the space a lack of sharp definition and an increased sense of turning inward.

The circular geometries that circumscribe both buildings are only a point of departure Saarinen seeks to resolve with a few modifications, to make them less – rather than more – "sharply defined." He therefore juxtaposes elemental[6] or *formal* objects with *informal* spaces within – not spaces that are casual, but that are intentionally irregular such that they cannot be summarized in a single Platonic solid or form. The circular geometries are contradicted to create dynamic spatial tension and reduce acoustical deadness – the irregularly sinusoidal brick waves, the eccentricities of floor steps, the inverted conical 'bowl' of the chapel ceiling that tapers asymmetrically, the contrasting curvatures beneath the auditorium ceiling, and so forth. In doing so, Saarinen provided clues of the external structure to the internal space, yet did not let the geometrical structures dictate the interior spatial configurations.

Furthermore, within their enclosure, the subordinate spatial divisions emphasize the difference in scale between the public realm in the auditorium and the corresponding public realm of the narthex in the chapel. Both admit entry from two sides with symmetric pairs of unsheltered doors,[7] but whereas the Auditorium beckons the eye upward to dynamically sloping planes beneath the seating above, the chapel presents a striking contrast of a tightly compressed, rectilinear narthex as a "decompression chamber" in Saarinen's terminology. His evocation of the parlance of modern travel emphasizes the tubular room's bridgelike connection, leading directly across the rhomboid grain of the marble floor tiles toward the eccentric internal steps.[8]

In both interiors, wood heightens the "informal" (or even "aformal") qualities not only through its material properties but as a material placed to curve alongside areas of human interaction, with varying and increasingly variable radii of curvature. Their variations increase or form a visual crescendo toward the focal areas. Saarinen shows a hierarchy between arc segments with fixed curvature and these splines with variable curvature: their curves appear to be chosen and expressive, which implies that they stand in implicit contrast to the exterior fixed geometry, which we shall revisit in considerations of the structure.

Therefore, at the auditorium and at the chapel (as in the TWA Flight Center and other works) Saarinen's apparent formal virtuosity with curves is not arbitrary and plastic [Saarinen 1962: 11], but is instead grounded by the hierarchy of axial relationships. That this is accomplished entirely without radial subdivisions is notable: Conventionally for the time, many other American modern architects incorporated centralized and circular geometries into contemporary religious and meeting structures. For example, in the Midwest, Frank Lloyd Wright's Annunciation Greek Orthodox Church (Wauwatosa, WI; 1956), Alden Dow's St. John's Lutheran Church (Midland, MI; 1953) [Robinson 1983], and John Randal McDonald's St. Therese Convent (Kenosha, WI; 1953) [Beno 1997] all use centralized forms and directed natural light to achieve intimacy within a larger congregational space. In contrast to Saarinen's consistent use of curvilinear geometry, Dow and McDonald introduce various polygonal geometries to rationalize the construction of the form. Wright and Dow place the altar in the geometric center, with seating in a circumferential pattern. These other structures are between the chapel and the auditorium in scale, so their degree of intimacy is somewhat diminished, but they make Saarinen's use of axiality rather than centrality that much more startling: since both the Chapel and Auditorium have no object or emphasis at the geometric center, one discovers their centers through acoustical reflections or deadening (where not to sit or play) rather than through material or spatial paths that heighten the center's importance.

2 Uses: Program and Experience

Along with the spatial reciprocity between the interior and exterior forms, there is a temporal tension between the use of the spaces as originally programmed and as experienced. These relate to differences of illumination quality and sound quality. While a full discussion of the programmatic goals of the MIT administration and possible unexecuted alternative schemes for both buildings would be somewhat conjectural, it is evident in the dedication that there was a burgeoning ambiguity as to how both spaces would be used primarily for intended programs of spoken lectures and 'non-prescribed' multidenominational worship.

First, both spaces seem to have been commissioned and envisioned as paired meeting places for an "academic community" that would also accommodate secondary performance uses. Just as MIT's application to the Kresge Foundation for a $1.5M grant noted that the auditorium should be "less distinctly religious in atmosphere" but "could be used for larger religious gatherings",[9] MIT President Killian's dedication stated that worship is first (for the auditorium, no less) and that music and drama linger further on:

> In visualizing an auditorium for MIT, we sought a building to provide a similar nucleus for our academic community, especially for our student body. We felt it would be proper and possible to design a building which would be appropriate for worship, for academic ceremonials, for educational meetings and conferences, for music and drama, and for the maintenance of the civil life of our academic community.[10]

Fig. 7. Kresge Auditorium, interior (view toward stage and organ loft).
Photograph by the author

The wood strips over plastic cloth that shield the acoustically absorptive glass-fiber blankets at the rear of the hall are the most visible acoustic measures, and they form an architectural transition between the reflective wood panels to the open wood screens at the transverse aisle (fig. 7). Saarinen's sectional drawings show the front movable portion of the stage configured lower as a seating or small theatrical 'orchestral pit' area, rather than having the movable portion elevated for an enlarged ensemble, since at design orchestra performance was intended to comprise perhaps 5% of the use.[11] Reporting in *Progressive Architecture*, George Sanderson observed the Holtkamp organs included in both the auditorium and chapel, but these permanent instruments, ancillary to the spaces,[12] were only the first indications of the role of music:

> [sidebar] Though the large 1238-seat auditorium was designed primarily for voice projection, it is also used for such musical events as organ recitals and symphony concerts.

[main text] On Dedication Day, we experienced the hearing environment under a wide variety of speeches, fanfares and processional marches by a Brass Choir; and the MIT symphony Orchestra, Glee Club, and Choral Society performed Aaron Copland's "Canticle of Freedom," especially composed for the occasion, and later in the program, a noble Bach cantata [Sanderson 1955].

The Copland piece, revived and performed by the MIT Wind Ensemble and Chamber Choir in 2000,[13] is not one of the composer's most known or lauded works, but the precedent had been set at its premiere: large orchestral performances would define much of the auditorium's role on campus, and MIT has still not erected a purpose-built music performance facility. As a result, without significant wing space and only the freight elevator connecting to the Little Theatre to move its large instruments, the space serves for rehearsals and performances of acoustic and amplified music from student and professional chamber quartets to international dance festivals.[14] Saarinen's landmark pair of structures serve as cultural sites within a campus possessing strong traditions of music performance in ways that have therefore become highly particular to both spaces.

The tension between the initial secondary role of music and its expanded role in the use of the Auditorium is mirrored in the chapel, where the initial primary intentions focused upon Saarinen's aforementioned "bilateral lighting" and its "unique" role in accommodating multiple religious traditions. Saarinen stated:

> The challenge of the interior was to create an atmosphere conducive to individual prayer. Since this is, uniquely, a non-denominational chapel, it was essential to create an atmosphere which was not derived from a particular religion, but from basic spiritual feelings. A dark interior seemed right – an interior completely separated from the outside world (to which the narthex passage would serve as a sort of decompression chamber). I have always remembered one night on my travels as a student when I sat in a mountain village in Sparta. There was bright moonlight over head and then there was a soft, hushed secondary light around the horizon. That sort of bilateral lighting seemed best to achieve this other-worldly sense. Thus the central light would come from the altar – dramatized by the shimmering golden screen by Harry Bertoia – and the secondary light would be light reflected up from the surrounding moat through the arches [Saarinen 1962: 36].

In this and other accounts, the role of the space for performance and for being a haven for long reverberant sounds are generally absent from the architectural accounts, save for the shaping of the interior wall for factual acoustic reasons rather than the expressive axial reasons stated above (figs. 8 and 9).

The actual impact of this powerful lighting scheme described earlier therefore varies with the actual worship practices in the chapel. As the definition of such a space as "multidenominational" has expanded from merely connoting Jewish, Protestant and Catholic in the 1950s, many of MIT's religious groups use the chapel for their respective worship times during the day and at night, and even more use the spaces in the adjacent Building W11 with its offices, meeting areas, tradition-specific food preparation areas, and other spaces absent from it.[15]

Fig. 8. MIT Chapel, screen detail (view upward toward ceiling detail at top of brick).
Photograph by the author

Fig. 9. MIT Chapel, interior lighting (showing light bouncing up through perimeter slot).
Photograph by the author

Some religious events do not occur within the chapel because its axial orientation is not compatible with their traditions, and others meet in spaces which are more "open" and less solemn than the chapel. In actual worship services, however, the incandescent lighting from the ceiling is almost always employed; natural light is almost always too dim to read comfortably in the space. Although the electric lighting is providing the actual usable light, the reflected light from the water and the overhead oculus remain lit in such a manner that the daylighting diagram remains evident while acknowledging the artificial light as secondary.

As the space is used for more than conventional religious observances and memorial services, its acoustics, which make a single voice able to dominate even when a hundred people are in the space, have also played a major role in defining it as a nonreligious performance venue. Its interior soundscape, with its acoustical properties similar to ecclesiastical buildings orders of magnitude larger, make it a compelling space for performances as well as for other student-initiated ceremonies. While its visual size makes chamber music of recent centuries seem appropriate, the persistence of sound and the delay between when multiple musicians hear each other indicate otherwise. The present Thursday Noon concerts offer the MIT community a range of musical genres, with a particular emphasis on early music suited for long reverberation times. As in the auditorium, the fittingly modernist pipe organ is a visually dominant but rarely used harbinger of the musical activities: the harpsichord or piano are used far more regularly, and a single tiny note can fill the volume. Thus, while performing in the space, qualities of liveness and of warmth to the resonance of a wide range of frequencies transform otherwise ordinary instruments' and voices' sounds into highly enriched and sustained echoes.

Fig. 10. Kresge Auditorium, exterior (view with temporary tents and Aalto's Baker House in rear). Photograph by the author

Therefore, while the design and the initial reception focused on the properties of vocal acoustics, in the decades since the acoustics have supported a broader range of instrumental uses; the importance of musical performance has illuminated how these two spaces have become small oases of natural materials, providing vibrant settings for student life within MIT's campus. The performance spaces may be within, but the periphery of the Auditorium plinth and the concrete edge of the moat serve as larger performative settings for campus social life: the brick steps form a rich outdoor setting with the curtain wall as a backdrop and its turf a stage for a variety of temporary pavilions such as a tensile fabric tents (fig. 10). These tents, stretched over the oval, take ideal shapes in their tensile material, their forms an inverted counterpoint to the compressive arches and domed shapes nearby. But how do Saarinen's iconic structural shapes communicate their structural logic?

3 Structure: Appearance and Reality

The dichotomies of human use and experience inform the third and most striking tension, between the appearances of purely compressive forms (masonry walls with arches, concrete shells), and the actualities of why and how they do not act in pure compression.

Saarinen states that the dome was chosen not only for its resonance with the other domes on the campus, but also for its efficiency as a modern form. Given uniform loading on the roof, the shape of a doubly-curved vaulted (rather than necessarily hemispherical) structure which is funicular and resists gravity loads in pure compression would approximate a catenary; the minimum radius of a catenary is less than that of a circle that passes through the same points, but the difference would be relatively slight given the shallowness of the vault. Saarinen recognized the impulse of structural integrity, but as with the initial catenary shape for the Gateway Arch in St Louis, his architectural articulation of the Auditorium form complicates its visible lucidity. The complications are not with the circular profile but with the manner in which it is supported and truncated: Domes and spherical shells can be easier geometries for engineers to calculate and builders to construct than those based on parabolas or other conic sections, and for a complete dome, the circumferential hoop action of the shallow dome-like shell contribute to its inherent stability and efficiency.

But since the one-eighth sphere is placed such that its outer perimeter is curved and ventures beyond the triangle inscribed by the three points of support these areas at the perimeter (shaded in fig. 6) are not acting in pure compression – transmitting loads directly to the three points of support – but are instead in tension. This means the shaded areas require substantial thickening of the shell overall and specific tensile reinforcement that contradicts the pure structural logic in order to maintain the external appearance.

> The dome is an equilateral, spherical triangle whose apexes reach down to the ground. There, they rest in pintle-like supports that are free to rotate. The structure is complex and, for a thin shell, quite thick. The structural, reinforced concrete part of the shell varies in thickness from 3.5 in. at the crown to 11 in. at the edges and over 20 in. at the points of the shell that sweep down to the supports [*Engineering News Report* 1963].

An ideal shell supported on the three points could be found experimentally using a hanging model or analytically using structural computations, but its form would be irregular rather than comprising conventional conic sections. Unlike other masterful compressive concrete shells by Felix Candela or those built more recently by Heinz Isler, designers have recognized that when Saarinen chopped the dome into a tricorner form, he destroyed the inherent balance he sought:

> The underside of the construction is a geometrically exact part of the surface of a sphere. Construction and form coincide. The hemisphere, out of which this form is cut, is statically in equilibrium under vertical loading. But, as soon as a piece is cut out of the hemisphere, the balance is destroyed. Additional forces have to operate at the edges to restore the shell's equilibrium. The distribution of these forces is not externally apparent in the construction of the auditorium. The form reflects the pure geometry of the sphere. As in other instances, the problem lay in deciding whether a shell ought to assume a form chosen beforehand, or whether the form should be dictated by the distribution of forces [*MIT buildings bibliography*: 126].

Fig. 11. Auditorium, Lobby (view south toward support condition). Photograph by the author

For Saarinen, the sculptural ideal of the spherical segmentation dominated over the tectonic ideal of other doubly-curved shell forms optimized to the points of support (fig. 11). It is now notable in retrospect that this prefigures the same challenges relative to the Sydney Opera House (which Saarinen championed as a competition juror), a design whose sketched shells were 'simplified' and modified into spherical geometries (long after Saarinen's death). These changes, imposing a single radius on each spherical segment, facilitated constructability and analysis by the team at Arup [Roman 2003; Jones 2006] (including the young engineer Peter Rice), but the result is vastly thickened and the shell functions structurally due to its thickened qualities rather than its shape. In a sense, all of the sculpturally shaped shells from Kresge onward through the Sydney example have demonstrated that even the prospect of a 'thin' shape creates numerous challenges in realizing the project:

'What's the philosophy behind a building like this?' I asked one who had to do with the auditorium. 'To enclose space at the least possible cost.' He replied. 'Did it?' 'Well, not perhaps this time, but from what we've learned if we had three to do each would be cheaper than its predecessor and the third would be considerably less expensive than the traditional form' [Weeks 1955].[16]

...[T]echnical problems ...arose as work proceeded, in spite of the most careful and expert analysis by Amman and Whitney, the engineers. ...[N]ot only has the structural analysis been thoroughly justified in practice but I have found the architect very willing to discuss his problems in most frank terms. Amendments to the design have been of a minor nature and were introduced as a result of contingencies which could not have been foreseen in the design stage of such a revolutionary structure, and they in no way detract from the appearance of the building [Scott 1955].

Yet the auditorium's shell is not a question of mathematical analysis, as these and other commenters seem to imply. Rather, it is a choice of conscious design manipulations: at MIT Saarinen set forth the spherical shell as an inevitable and efficient structural solution of supreme priority, placing all interior acoustical and construction modifications as inherently secondary to the structural motivation. Again, to continue the above assessment from 1955 by architect, planner, and founder of the BDP music society N. Keith Scott:

There can be little doubt that the architect's departure from the logical approach to design, as we currently understand that term, has created many structural and mechanical problems, and in many instances he has relied upon the sheer ingenuity of modern technology to get him out of difficulties...In his choice of structural form Saarinen flouts every precept of basic acoustical design, for the concave ceiling and the curved rear wall combine to prohibit good hearing conditions unless there is a vigorous appliqué design to counter the tendency to focus sound [Scott 1955].

Nonetheless, Saarinen remains steadfast in his logic:

One concept underlying auditorium design is to let the functional requirements – acoustics and sightlines – determine the form. But there is no one ideal acoustical shape. Though function has to be respected, it seemed equally justifiable to let the basic form come from structure. Thus, in developing the design of this building, we felt very strongly guided by Mies' principles of architecture – of a consistent structure and a forthright expression of that structure. As there are many ways of doing equally functional things, we built dozens of models. As an auditorium requires a triangular shape, we tried spanning this one-room building with a dome supported at three points – the shape of one-eighth of an orange. At first it seemed strange, but gradually it became the loved one [Saarinen 1962: 34].

Therein lies Saarinen's shell game: the sphere persisted in the guise of a purely efficient solution, even though this purity was no longer an ideal form for the truncated plan shape or for other interior aspects.

Nevertheless, the descriptive elegance of the one-eighth orange, melon, or other handy illustrative fruit from the apocryphal story has persisted in historical discourse. In the years since shells have no longer been economically efficient since expenses of labor and formwork have outpaced material conservation in many Western economies. Practitioners and students who study the auditorium rarely anticipate or comprehend the structural gymnastics rendered invisible from within and without.

At a much more subtle scale, the brick arches of the MIT chapel have similar hidden inner workings: While arches effectively support planar brick walls, transmitting their distributed loads to discrete points of support, the curvature in plan of the cylindrical face means that in perimeter the brick swerves beyond the line of action of the arch.[17] This eccentricity invites viewers to speculate as to how 'true' the brick is. Saarinen may have stated that this is undeniably a masonry structure –

> It seemed right to use a traditional material, such as brick, for the chapel – for brick would be a contrast to the auditorium and yet the same material as the surrounding dormitories. But we felt that brick should be used with the same principles of integrity to material as concrete or steel. This is forthrightly a brick structure [Saarinen 1962: 36] –

but in truth this is an expression of structure rather than a diagram of its performance, as the visible brick surfaces enclose the layered assembly within.

> The arches on the outside occur where the exterior wall and the undulating interior wall meet. In retrospect, especially having looked again at the archivolts of Romanesque churches, I wish that we had given these arches a richer, stronger three-dimensional quality. And I am aware that the connection between narthex and chapel is clumsy. However, I am happy with the interior of the chapel. I think we managed to make it a place where an individual can contemplate things larger than himself [Saarinen 1962: 36].

While no other permanent shells have been built at MIT since 1955, issues of shell geometry have been a recurrent leitmotif of design exploration: architectural faculty members such as Eduardo Catalano were famous for their shells, such as the hyperbolic paraboloid for a house in North Carolina. Longtime faculty member Waclaw Zalewski built dozens of highly efficient thin-shell roofs for factories and other spaces in his native Poland and in Venezuela before teaching at MIT from 1966-1988.[18] Several generations of designers in the MIT community have collaborated on recent educational exhibits and publications that seek to communicate the usefulness of graphical approaches for designing vaults, domes, and even more conventional trusses, walls, and columns. These educational aims, used in teaching at MIT and by associated faculty members practicing and teaching far abroad are captured in *Form and Forces* [Boston Structures Group 2009], a book that provides instruction in the processes of graphic statics as well as form optimization for all major structural systems, and uses works such as Saarinen's main terminal at Washington Dulles International Airport to illustrate applications of structural design strategies that use the shape of structures for strength and efficiency rather than for arbitrary purposes. Saarinen's works do in fact demonstrate effective structural logic. Saarinen himself was aware of these dichotomies. In an address given at Dickinson College on December 1, 1959, he said:

> The principle of structure has moved in a curious way over this century from being 'structural honesty' to 'expression of structure' and finally to

'structural expressionism.' Structural integrity is a potent and lasting principle and I would never want to get far away from it. To express structure, however, is not an end in itself. It is only when structure can contribute to the total and to the other principles that it is important [Saarinen 1962: 6].

And in a letter to a friend dated June 3, 1953:

Esthetically, we have an urge to soar great distances with our new materials and to reach upward and outward. In a way, this is man's desire to conquer gravity. All the time one works, one concerns oneself with the fight against gravity. Everything tends to be too heavy and downward pressing unless one really works at it [Saarinen 1962: 5].

Postlude: Performing and Hearing

Finally, the acoustic qualities that result from and within Saarinen's shells, vaults, and walls are among the most iconic of American modernism. While scholars such as Emily Thompson [2002: 319-320 ff] have articulated the rich evolution of the "modern sound" for controlled acoustics in the earlier decades of the twentieth century, Thompson rightly foreshadows how changing tastes and preferences motivated consultants such as Bolt, Beranek, and Newman to reach beyond reverberation time toward a more qualitative assessment of parameters such as the "initial time-delay gap" that linked data and perceptive opinions. Leo Beranek's book *Music, Acoustics, & Architecture* focused on a litany of eighteen qualities that support excellence in concert hall acoustics, and reverberation time is *not* one of them although it plays a constituent part in half [Beranek 1962].

Fig. 12. Chapel, Interior (Detail at perimeter). Photograph by the author

Fig. 13. Auditorium, Lobby Ceiling and curtain wall. Photograph by the author

The single most important quality, according to Beranek, is intimacy, followed by (in combination) liveness, warmth, and the loudness of direct sound and loudness of the reverberant sound, and then (in combination) diffusion, balance, and blend.[19] These qualities relate specific acoustic phenomena to a range of geometrical and material connections, and translate them into the language of musicians. For example, the relative audibility of bass-range sounds improves warmth, while intimacy is measured by the shortness of time (the duration or "gap" mentioned above) between the initial sounds and the first reflected sounds as determined by the proportions and physical geometry of the surfaces that bound a space. That these experts collaborated with Saarinen for the auditorium indicates that its acoustics should be viewed as a transitional experiment leading up to the major works Beranek and his colleagues collaborated upon, and an example of a building notable in that the reverberation time is not what characterizes the success at the auditorium for various performing forces.

> When the Kresge building was in its next-to-final stage members of the Institute faculty used to wander into the main auditorium and standing in the pit clap their hands for the joy of hearing the sound bounce from the cement floor to the cement dome and back again; then they would seek out Mr. Newman and tease him. But the acoustical engineers had the last laugh; with delicate wooden gratings, with unobtrusive backstops of plastic fabric hung over fiberglass pads, with sound-absorbent seats, they have controlled and clarified the voices that come from the stage…[who] spoke well and were heard with ease. The single voice and the solo instrument are beautifully accentuated, but it is still a question of how true a blend we shall get from the full orchestra [Weeks 1955].

Just as the roof membrane and metal panels over the shell were substantially replaced not long after construction, the interior of the Auditorium has also had ongoing acoustical tweaks as visible in the configuration of the "clouds" to become more undulating and more extensive. In its 'deliberate reach for shape, beyond the definition of function' that made it "a pivotal one in US architecture today" in 1955 and a half-century later, Saarinen created structures for creating community and introspection, but the new faithful that come to venerate are the architects themselves:

> Fifty years after the construction of Kresge and the chapel, architects still make pilgrimages with their sketchbooks and cameras, and try to figure out how this master of mid-century modernism did so much with seemingly so little.[20]

The acoustical tensions and dichotomies between the spaces as perceived visually and aurally serve to reconcile the internal complexities with the overarching gestures which makes Saarinen's works such resonant contributions to American modernisms. That they have not only iconic imagery but iconic sounds is further evidence of how they transcend the limitations of forms that could be static. Saarinen's work here and elsewhere constitutes not a series of built diagrams – simple forms reduced to their least complicated – but rather a group of richly particular humanistic experiments that negotiate the architectural pursuits of external clarity and internal discovery. Saarinen places these in dynamic tension such that the act of listening within the spaces – however unexpected and atypical – is just as particular and unreplicated as the forms and lighting effects. Echoing the common theme in religious traditions continuing in the chapel, the architecture and the soundscape of the Auditorium are in their world but not of it; they recognize their role in the urban campus to transcend their context and create independent environments for listening. Therefore, the visible architecture is not an end but rather a means towards creating a more total sensory environment for spiritual and temporal meaning. In Saarinen's words:

> I look for the day when our spiritual qualities catch up with our physical advances. Then our architecture will take an important place in history... [Peter 1994: 193].

Notes

1. The chapel design at MIT contrasts Saarinen's chapel at the Lutheran Seminary in Ft. Wayne, IN. In this instance, Saarinen designed not only the chapel and its environs, but the residences and other surrounding campus buildings. That chapel, however, is a large, pointed form based on an abstracted stave church, relating to the Lutheran church's Scandinavian heritage. Unlike the brick chapel at MIT, the example in Ft. Wayne is constructed mostly of concrete, yet it shares the strip of horizontal windows that admit light into the nave from a shallow outdoor pool. Due to the lighter tone of the concrete and the angled walls, the author observes that light is reflected somewhat more readily into the space than at MIT, but without the textural surface articulation achieved with brick.
2. The wartime growth of MIT had brought in many commuter students, and there was a large parking lot for a time. After World War II, the development of student life and humanities programs as described in the 1949-50 MIT Lewis Report coincided with the construction of MIT's first dorm west of Massachusetts Avenue.
3. "The chapel's spire and bell tower, designed by sculptor Theodore Roszak, were added in 1956. The bell, also designed by Roszak, was cast at the MIT foundry, which was then located on the top floor of Building 35" [Wright 2005].
4. "Seen by night, with its lights within, it is an opal; by day it suggest one of those curved white hats the ladies have been affecting this spring, or, less elegantly and to the critics, it suggests a diaper. Since the married students are housed directly behind the Auditorium it might be that Mr. Saarinen had this symbol in mind, though I doubt it" [Weeks 1955].

5. "Kresge Auditorium at MIT has been completely reroofed just eight years after it was opened. The surface of the auditorium's remarkable three-point supported, concrete dome roof was in trouble almost from the beginning. MIT's maintenance engineers, who incidentally collaborated on the original design, watched in dismay as violent and unexpected thermal stresses weakened and destroyed the outer layer of the shell's three layer system" [*Engineering News Report* 1963].

6. "In designing a chapel suitable for use by worshipers of all faiths at the Massachusetts Institute of Technology, Eero Saarinen chose as his starting point the elemental cylindrical form. The way he made the form live without destroying its essential simplicity reflects his careful consideration of every part" [Grubiak 2005].

7. See comments in [Weeks 1955].

8. One could compare the form of these in plan to conventional depictions of moving sound sources and Doppler waves, but that would be highly conjectural and is not supported by Saarinen's documentation.

9. See [Grubiak 2007], note 33, "Submitted application to establish at MIT a Kresge School of Human Relations," 11 April 1950, AC 4 MIT Office of the President, 1930-1958, box 131, folder 12, MIT Archives.

10. See also [Grubiak 2007], note 32: "Full text of an address prepared by Dr. James R. Killian, Jr., President of the Massachusetts Institute of Technology, for delivery at the Dedication of the Kresge Auditorium and the MIT Chapel at 3:30 o'clock on Sunday afternoon, May 8" with annotations, AC 4 MIT Office of the President, 1930-1958, box 131, folder 8, MIT Archives.

11. "50% for lectures and drama, 15% for organ, 5% for orchestra, 18% for chamber music and operetta, 12% for soloists and glee club" [Beranek 1962: 109]. In the past decade, the relative proportion of orchestral and large chamber rehearsals and performances has exceeded all other uses, perhaps even large but infrequent conference lectures.

12. "Most organists point out that the reverberation time is too low..." [Beranek 1962: 108].

13. December 2000, Kresge Auditorium. The author played in the MIT Wind Ensemble for this performance.

14. MIT has a music faculty larger than most other non-conservatory institutions in Boston, serving over a thousand students who participate in music courses and ensembles each year; the minor course of study in music is among the largest in the Institute, and MIT music majors, such as Alan Pierson (Alarm Will Sound chamber orchestra), Andrew McPherson (Tanglewood composition fellow), and others, have become nationally recognized composers, conductors and performers.

15. The chapel was envisioned in some Saarinen studies to have a stronger rear bar with library and office support functions, a version published but not executed; the rear wall simply lines the alleyway with a single mute egress door; the masonry itself had deteriorated over the length of the wall and its southern end was rebuilt in 2007-8.

16. Weeks also muses: "If in the intermission, there should be a choice between continuing with a difficult pay or concert or going home to an open fire, I wonder whether the fire would not win."

17. Special thanks to Edward Allen FAIA for his illumination of this related point and for his helpful commentary on structural design issues more broadly

18. In retirement, Zalewski has collaborated and co-taught with former faculty such as Edward Allen FAIA, who along with MIT's architecture, building technology, and engineering faculty and students have explored the analysis and design of compressive structures with newfound computational tools as well as centuries-old practices of graphic statics. See also http://www.shapingstructure.com and http://web.mit.edu/masonry.

19. Cf. [Nuzum 2009]. Beranek's eighteen qualities are: [intimacy (presence), liveness, warmth, loudness of the direct sound, loudness of the reverberant sound, definition (clarity), brilliance, diffusion, balance, blend, ensemble, immediacy of response (attack), texture, freedom from echo, freedom from noise, dynamic range, tonal quality, and uniformity. Even though Beranek developed an interrelated way to use these qualities to 'score' concert halls, Nuzum notes that the manner by which the eighteen qualities are inextricably linked to each other means that at least nine of them relate directly to reverberation, further evidence of the qualitative specificity

of Beranek's descriptions: "Just as the Inuit people of the northern reaches of North America are said to have hundreds of words for all the different kinds of snow, in the world of acoustics, reverberation is known by many names" [Nuzum 2009: 10].

20. William J. Mitchell, professor of architecture and media arts and sciences, quoted in [Wright 2005].

References

ALBRECHT, Donald, ed. 2006. *Eero Saarinen: Shaping the Future*. New Haven: Yale University Press, in association with the Finnish Cultural Institute in New York.

BENO, Fr. Brian. 1997. *John Randal McDonald*. Milwaukee, WI: Private Printing.

BERANEK, Leo. 1962. *Music, Acoustics, & Architecture*. New York: Wiley

BOSTON STRUCTURES GROUP (Allen, Zalewski, Foxe, Anderson, Ochsendorf, Ramage, et al.). 2009. *Form and Forces: Designing Efficient, Expressive Structures*. New York: Wiley.

DELONG, David G. 2008. *Eero Saarinen. Buildings from the Balthazar Korab Archive*. Library of Congress visual sourcebooks in architecture, design, and engineering. New York: W. W. Norton in association with the Library of Congress.

Engineering News Record. 1954. Tripod dome built on tricky formwork. *Engineering News Report*, 27 May 1954.

———. 1963. Roof Repair: More than Skin Deep. *Engineering News Report*, August 8, 1963.

FOXE, David M. 1999. Finland: Architecture and History – The quest for national identity from traditional vernacular architecture through Saarinen and Aalto. Unpublished.

GRUBIAK, Margaret M. 2007. Educating the Moral Scientist – the chapels at IIT and MIT. *ARRIS The Journal of the Southeast Chapter of the Society of Architectural Historians* 18: 1-14.

JONES, Peter. 2006. *Ove Arup: Master Builder of the Twentieth Century*. New York: Yale University Press.

MIT buildings bibliography. Cambridge, MA: Rotch Library, Massachusetts Institute of Technology, 198-?]-T171.M423.R68 1980

MIT Committee on the Visual Arts. 1988. *Art and Architecture at MIT: A Walking Tour of the Campus*. Cambridge, MA: MIT Press.

MITCHELL, William J. 2007. *Imagining MIT: Designing a Campus for the Twenty-First Century*. Cambridge, MA: MIT Press.

MERKEL, Jayne. 2005. *Eero Saarinen*. London; New York: Phaidon.

NUZUM, James M. 2009. How and Why Architects Should Claim Acoustics. Unpublished research conducted for a 2009 course "Variations in Sound and Architecture" at the Boston Architectural College taught by the author of this paper.

PETER, John. 1994. *The Oral History of Modern Architecture*. New York: Abrams Publishing.

ROBINSON, Sidney K. 1983. *The Architecture of Alden B. Dow*. Detroit: Wayne State University Press.

ROMÁN, Antonio. 2003. *Eero Saarinen: An Architecture of Multiplicity*. New York: Princeton Architectural Press.

SAARINEN, Eero. 1962. *Eero Saarinen on his Work: A Selection of Buildings Dating from 1947 to 1964 with Statements by the Architect*. New Haven: Yale University Press.

SANDERSON, George. 1955. New MIT Buildings Open. *Progressive Architecture* 36, 10 (July 1955): 74-75.

TEMKO, Allan. 1962. *Eero Saarinen*. New York: G. Braziller.

SCOTT, N. Keith. 1955. MIT auditorium: an English view. *Architectural Record* 117 (July 1955): 138.

THOMPSON, Emily. 2002. *The Soundscape of Modernity: Architectural Acoustics and the Culture of Listening in America, 1900-1933*. Cambridge, MA: MIT Press, 2002.

WEEKS, Edward. 1955. The Opal on the Charles. *Architectural Record* 117 (July 1955): 131-137.

WRIGHT, Sarah H. 2005. Architectural wonders, chapel, Kresge turn 50. *MIT News Office*, October 19, 2005. http://web.mit.edu/newsoffice/2005/chapel-1019.html.

About the author

David M. Foxe holds degrees in architecture and in music from MIT, and was a Marshall Scholar to Clare College, Cambridge (UK). A former architectural tour guide of the two Saarinen structures as well as Aalto's Baker House at MIT, he draws upon experience as a music composer and performer who has written and played vocal and instrumental music in both spaces since 1999. While continuing to teach courses and publish works on design history, structural techniques, and the relationship of architecture and music, he currently practices architecture in Boston and practices piano in Newton.

Ozayr Saloojee

University of Minneapolis
College of Design
School of Architecture
UMN Twin Cities
89 Church St S E
Minneapolis, MN 55455 1
saloojee@umn.edu

Keywords: Eliel Saarinen, Eero
Saarinen, Christ Church
Lutheran, modern architecture

Research

The Next Largest Thing: The Spatial Dimensions of Liturgy in Eliel and Eero Saarinen's Christ Church Lutheran, Minneapolis

Abstract. Christ Church Lutheran in Minneapolis, Minnesota, was designed by Eliel Saarinen, then 75, and added to by his son Eero Saarinen 10 years later. Deeply loved by its community, it also serves as a touching example of the relationship between the father and the son. This present examination looks at the building on various scales, underscoring the finesse and material elegance of the building complex, the spatial genius and expertise of Eliel Saarinen, and the deferential addition by Eero.

Written into the National Register of Historic Places, Eliel and Eero Saarinen's Christ Church Lutheran was named as a National Historic Landmark by the U.S. Department of the Interior in early 2009. Widely considered to be Eliel Saarinen's masterwork, the church has been hailed as the singular building example that heralded a new direction for ecclesiastic architecture in the United States. Completed in 1949, Eliel's sanctuary sits – in the words of his grand-daughter Susan Saarinen – "quietly there".[1] It is an unprepossessing building from the exterior, a massing of simple forms – largely rectangular seeming solids, faced with Chicago brick and Mankato Stone, occupying the corner of 34th Avenue South and East 33rd Street in Minneapolis's Longfellow neighborhood. Its exterior belies its interior, which demonstrates Eliel Saarinen's consummate skill as an architect capable of understanding the scale of experience as an essential part of liturgy and as an evocative catalyst for a deep and personal sense of spirit. Approached by the local congregation shortly after World War II, Pastor William Beuge wrote a challenging letter to Saarinen, then head of the Cranbrook Academy of Art in Ann Arbor: "I asked him if it were possible in a materialistic age like ours to do something truly spiritual." The young pastor observed that, "He soon showed me".[2]

An important question during this ongoing research into Christ Church Lutheran has regarded how to approach a reading of this building. This research – and this writing – is not the work of an historian, so the question of how to uncover or tease out readings of this building has been a preoccupying concern; it has perhaps resulted in an initial reading that departs from an expected analysis or interpretation. Compounding that is the focus on a building that has been the subject of numerous, if not hundreds, of interpretations over its almost sixty year history. Christ Church Lutheran has been studied and drawn by students and architects from all over the world, recently hosting part of an international symposium paralleled with the exhibit "Eero Saarinen: Shaping the Future", mounted in Minneapolis at the end of 2009. Curiously, however, the building is not as firmly embodied in the architectural radar of those who visit the area; building aficionados are more likely to know Jean Nouvel's (new) Guthrie theatre (and sadly, be less familiar with Ralph Rapson's original – now torn down), or Jacque Herzog's and Pierre de Meuron's Walker Art Center, or the Gehry project – the Weisman Museum –more than the Saarinen project.

Nexus Network Journal 12 (2010) 213–237
DOI 10.1007/s00004-010-0032-6; *published online* 11 May 2010

Fig. 1. Eliel Saarinen helping lay the cornerstone of Christ Church Lutheran in 1949. Image © Christ Church Lutheran Archives. Still from archival footage digitized by author

Fig. 2. Eliel and Loja Saarinen at the building opening. Image © Christ Church Lutheran Archives. Still from archival footage digitized by author

That Christ Church Lutheran was designed by Eliel Saarinen at the age of 75, and added to by Eero Saarinen 10 years later, adds to the knotty question of how to approach an interpretation of this particular building. Eero was asked by the congregation in the early 1950s to design the addition to his father's building. Two very distinct and idiosyncratic masters shaped this project over a period of more than a decade. Apart from all of this, perhaps the most important aspect that has been brought to the fore in the study of this building is knowing that it is, and has been, the object of a deeply felt love by its community – well before the cornerstone was laid in 1949 (figs. 1, 2) – and that it serves as a touching example of the relationship between the father and the son.

If architects design for effect and to affect, then a mark of the good architect is to do so without compromising either the technical or the experiential. This dual capacity of the good architect – the ability and nuance to negotiate the scale of experience between the mechanical and experiential – was one of Eliel Saarinen's great skills. In considering that the building was authored both in the name of the father and the son, this notion of "effect" (and in a way, of "cause") has a larger resonance; the hand of two designers with markedly different attitudes about form and space underpin this building.

The evidence of this is embodied at Christ Church Lutheran in two particular ways: in the subtle deftness of Eero's addition to his father's building, and in Eliel's designed experience of the sanctuary. Eliel Saarinen was a master of scale: his work encompasses the broad scope from the local to the global. He designed and proposed city plans all over the world: in Estonia, Finland and Australia, even consulting on projects in South America. He designed buildings – museums, apartment, manors and campuses – and he designed furniture and fixtures – from chairs to tables to teapots and teacups.

If the normative way to look at a building would be from the outside in, this investigation will start instead from the inside out – from object to space to room to building. By looking at these scales, and through these lenses, an attempt is made to underscore the finesse and material elegance of this building and to touch on the spatial genius and expertise of Eliel Saarinen, and a surprisingly soft and deferential project by Eero.

Eliel Saarinen was never satisfied with acceptable, or even good responses; he attempted to conceive of projects in their entirety, at all scales and, perhaps most importantly, to know what the appropriate use of these scales are. The range of scales is embedded here at Christ Church Lutheran, and it is reflected in many ways: from the material and tectonic to the immaterial and the ineffable; to the way that choral music surrounds worshippers and visitors; to how a monumental experience of space is made personal and intimate; to the subtle and deeply poetic way in which light is treated, used and shaped. Eliel understood how light becomes a material principal of this building. In the Lutheran tradition, God's Grace comes by faith alone, through Christ alone – *Sola Gratia, Sola Fide, Solus Christus*. Light, of course is one of the mechanisms with which architects and designers try to express a sense of the sacred. At Christ Church Lutheran, for Eliel Saarinen, light revolves around liturgy connotatively and denotatively.

Eliel Saarinen would have been very familiar with this liturgy. His father, Juho Saarinen, was himself a Lutheran minister, tending to congregations in Rantasalmi, Finland and in Lisisila in Ingermanland, Russia. Juho would later leave his congregation in St. Petersburg and move to the outskirts of Helskini, to join his son and family at their Hvitträsk estate. The Lutheran liturgy is organized by four particular movements:

- Gathering;
- Word;
- Meal;
- Sending.

These movements are embedded at varying scales in Christ Church Lutheran, and overlap through both the material and immaterial qualities of the building. They serve as a kind of conceptual touchstone for the reading of this building and as a narrative glue. This interpretation will, as noted earlier, start with an inversion of sorts, moving from the micro and shifting out towards the macro, a kind of Eamesian "Powers of 10," looking at both Christ Church Lutheran and its designers, Eliel and Eero Saarinen.

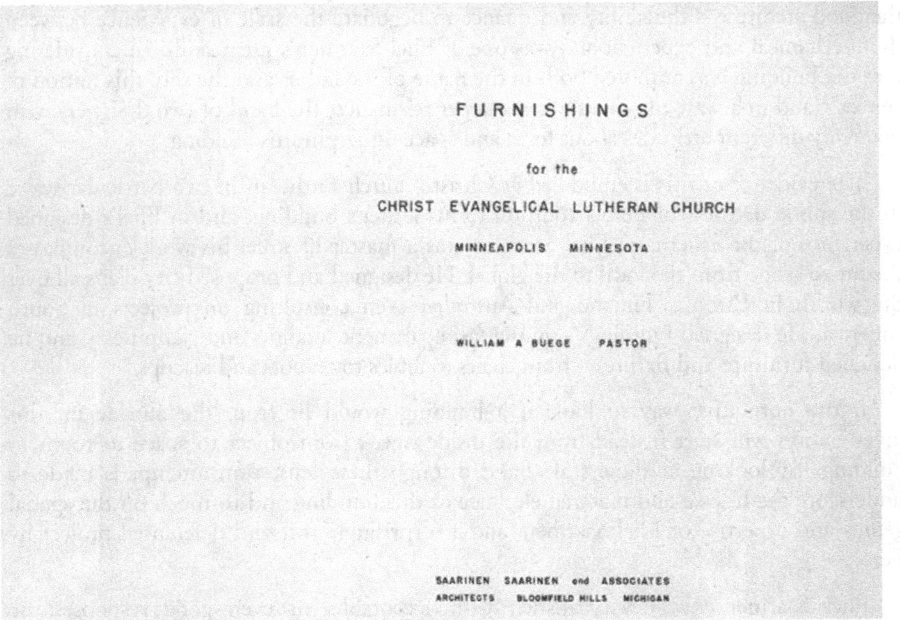

FURNISHINGS

for the

CHRIST EVANGELICAL LUTHERAN CHURCH

MINNEAPOLIS   MINNESOTA

WILLIAM A BUEGE   PASTOR

SAARINEN SAARINEN and ASSOCIATES
ARCHITECTS   BLOOMFIELD HILLS   MICHIGAN

Fig. 3. Title-page, 'Furnishings Book' for Christ Church Lutheran.
Image © Christ Church Lutheran Archives

Appropriately, for a church, we can begin with a book. Not THE book, but a book (fig. 3), and certainly an important one; The Book of Details, if you will, and a gospel for an architectural understanding of Christ Church Lutheran. Provided for the church as part of the final set of construction drawings for the building, this book contains original drawings from Saarinen's office showing the furnishings for the church. If we start at the scale of the liturgical objects depicted within, we can extract aspects from them that allow us to posit a manner in which the architect understood his work, not just in the form of the building, but also the way in which he communicated these ideas to us and the community at large. The drawings in this book parallel the way in which the represented objects themselves exist in the physical space of the church, how they occupy it and are highlighted by it. This is a key relationship. British artist David Hockney observed that how we depict space indicates how we behave in space;[3] there is a relationship between the image of the object and our experience of it. Here, in the good Book of Details, we see these representations – these object-drawings – as artifacts displayed against a surface, organized against the backdrop of the page. Their drama is played out not only in the usual convention of the architectural drafting board, through plan, section or elevation, but also in perspective, with cast shadows. These objects have been rendered as experienced things. The images in the book can therefore be understood as an expression of architectural thinking through drawing; before the objects were made, they were imagined as matter, thought of as matter. These liturgical artifacts themselves can then be understood as the manifestation of this thinking through making/drawing. At Christ Church Lutheran, the immaterial was understood through the material realization of ideation.

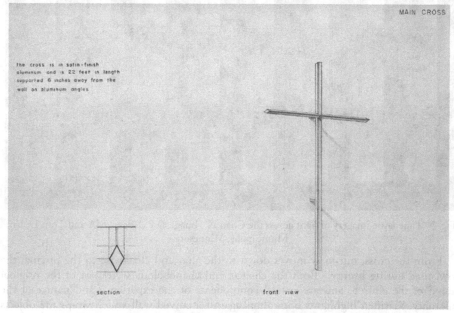

Fig. 4. Main Cross, 'Furnishings Book' for Christ Church Lutheran.
Image © Christ Church Lutheran Archives

The cross (fig. 4) is the focal point of the chancel and for the congregants as they gather, in fulfillment of the first movement of the Lutheran liturgy. The volume of the sanctuary itself and the subtle geometry of the nave walls direct one's vision forward. The cross – the ultimate symbol of Christianity, representing at once Christ's saving death and His resurrection – casts an ever-lengthening shadow along the rear of the chancel wall. Through the architecture, the artifact sits against a backdrop of light falling upon the Chancel wall: a material scrim for the divine.

The cross is 22 ft. tall, constructed of stainless steel and mounted to the wall with aluminum brackets. Its section is diamond shaped, with clear, hard edges. Recollections from Nick Hayes, the son of Mark Hayes of Hills, Gilbertson and Hayes, the local office associated with the Saarinens, tells the story of one of Eliel Saarinen's periodic site visits. Laborers had painted the chancel wall with two coats of white, oil-based paint, and Saarinen, upon seeing it, instructed them to grind back the surface of the newly painted Chancel wall, knowing that the softened backdrop would yield a more appropriate experience of the cross and its shadow. Indeed, the drawing of the cross in Saarinen's book of details articulates the dramatic intentionality of the liturgical object against an ephemeral backdrop; the shadows are rendered against an implied surface. In constructing the effect of the cross, Saarinen simultaneously understood its affect (fig. 5).

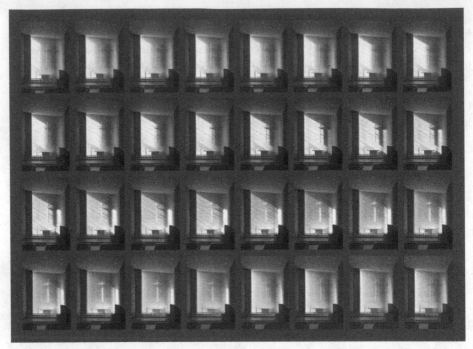

Fig. 5. Time-lapse imagery of light across the Chancel. Image © P. Sieger, AIA and Tom Dolan, Minneapolis, Minnesota

From the cross, our view moves down to the altar and clockwise to the piscina, the bowl used during liturgy. Both the chancel and chapel altars were part of the original design of the church, and are critical components of the experience and nature of the sanctuary. Saarinen highlights both altars against a curved wall which wraps the objects with the diffuse light that enters through translucent windows. The chancel wall curves forward from the sanctuary altar and leads to the piscina, which is located near the altar used for Communion and is mounted on the same wall as the cross. This curve leads to the pulpit, which has its own curved geometries that align with the side-aisle to the north. This suggests a line towards the baptismal font – completing a kind of liturgical loop – one that highlights the space of the Chancel, strongly evident in the plan of the building. The pulpit (fig. 6) is the station of the utterance of the Word – the second movement of the Lutheran liturgical movement – and is in a way a centralizing moment within the space itself.

Through these objects, Saarinen observes and celebrates the integrity of a set of individual moments within the space, but this isn't a hermetic strategy because each of these spaces is impacted by the broader program of the building. These object-moments exist as individual events, but are connected, both conceptually and experientially to the larger scale of space around them.

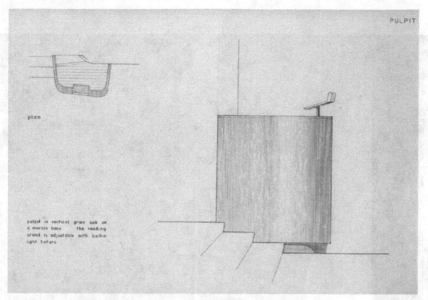

plan

pulpit is vertical grain oak on
a marble base   the reading
stand is adjustable with built-in
light fixture

Fig. 6. Pulpit, 'Furnishings Book' for Christ Church Lutheran.
Image © Christ Church Lutheran Archives

The altars and their candles are also objects against the page (as drawn in the book) but are also projected against the walls of the church, or in the case of the baptismal font, against a volume defined by its location: situated in a curve in the floor, in a space of its own, with a flooring pattern that differs from the surfaces around it. The Baptistery is bounded by a short, low wall and one steps down into it. The floor is a glossy tile, distinct from the other material surfaces of the sanctuary. Stepped down, it recalls Christ's baptism in the River Jordan by John the Baptist.

These liturgical artifacts – the cross, the altars, the piscina and the font – have deliberately considered and architecturally nuanced spatial and experiential conditions. As we expand our scale beyond the object as a single moment, we can observe that they are always inserted in a larger, careful material and luminous context. Although the space of the liturgy occupies the entire spatial field of the sanctuary – indeed, the whole building – Saarinen draws out a set of careful architectural conditions that highlight these liturgical and (to coin a term) artifactual moments. The curve of the chancel wall (fig. 7), and the wooden baffle or screen carefully modulate the light that falls onto the surface of the whitewashed brick. All the glass used in the sanctuary (save the entry and bell tower windows) is translucent, which further tempers the eastern and southern light; Eero Saarinen would later use this exact strategy in his design for the Kresge Chapel on the M.I.T campus in Cambridge, Massachusetts. Specific walls are terminated in such a way that the visible cut edge of the walls themselves define program and space within the building. For example, the two flanking walls from the narthex to the sanctuary are faced with brick headers only, as is the short wall of the baptistery. These edges help articulate particular zones or programs of the church in a less explicit way. The chancel wall edge, perpendicular to the window, has its own pattern of headers and stretchers; further emphasizing this particular volume and its bounding by the symbol of the cross, the station of the Word, the font and altar assist in expressing it as a space of elemental spirituality (fig. 8).

Fig. 7. Sanctuary, view of the chancel and cross. Image © P. Sieger, AIA and Tom Dolan, Minneapolis, Minnesota

Fig. 8. Sanctuary. Image © P. Sieger, AIA and Tom Dolan, Minneapolis, Minnesota

Saarinen is deliberate and methodical about the walls in Christ Church, the way they extend, connect, imply or are cut and exposed. The west wall that bounds the baptistery extends outward, well past where we would expect it to merge with the southern aisle wall. In doing so, it mimics a similar strategy used in the Chapel's East wall. The material and formal expression of these moves point to a considered hierarchy of the spaces of the church; Saarinen's attitude toward the building is clear, in that his architecture underscores spaces of profound spiritual experience within the building, but he does so with quiet poise. The side aisles, too, use this strategy: the west aisle links to the heavy volume of the bell-tower, the east connects to the narthex and lobby. Both connect the glazed end of an aisle wall to an opaque masonry volume.

For those coming for the first time, the reaction to the building is often one of casual indifference. Certainly from the outside, the Saarinen complex maintains a remarkably unassuming and unprepossessing presence. Upon entering the building however, that casual indifference changes quickly. The church is raised up a few steps on a plinth above the level of the sidewalk, and one enters an arcade before turning to the main doors of the building. The arcade was not part of Eliel's early design, but was instead added through Eero's addition. Eliel Saarinen, with the entryway, demonstrates not only his careful attention to detail with a carefully and intricately crafted door, but also in framing the entry with a mullion-less glazing, clearly separating two formal elements within the entry spaces of the church, again suggesting the integrity of these moments within the overall structure and experience of the architecture. He then does this in several places within the building and the result is anything but fragmentary; he achieves a kind of architectural unity through multiplicity of scale. He also does this at the campanile, an object linked by glass on both sides, and tied into the structure of the building through roof, floor and stair. It reads simultaneously as an artifact in its own right, against the backdrop of the volume of the building of the church, but the stairs that lead to the choir loft are pulled off the side walls by a small gap of a few inches, which allows light to penetrate between wall and stair. This scale of separation – at the scale of both the detail and the volume of the building – would lead Eero Saarinen to employ a similar strategy in the double stair of the addition that leads down into the basement level of the education wing.

Fig. 9. Side-aisles. Image © P. Sieger, AIA and Tom Dolan, Minneapolis, Minnesota

The movements of the Lutheran liturgy are ancient, and go back to the very beginnings of the Christian Church. The entry of the congregant into the space is directed forward to the altar and the cross and the chancel wall; this is made more emphatic by the volume of the sanctuary and smaller coves of the side-aisles. The side-aisle windows are angled forward, with a smaller edge profile that reduces the amount of visible lit surface as the congregants gather. The second movement of the liturgy is the Word – *Sola Scriptura* – which takes the pulpit as the seat of the book, and shifts the focus of the congregants to the pulpit from the cross. The Meal directs worshippers forward again to the altar. The manner in which the third movement is performed underscores how Saarinen's architecture augments the experience of the Lutheran liturgy and how this building can be understood as a facilitator for worship. Once complete, congregants split into two lines and move back to the pews, along the side-aisles. This time, however, they are exposed to the broader face of the side-aisle windows, with their edges more prominently in light (fig. 9). The view back, towards the east, is again, a brick surface, softly lit from either side through translucent glass.

This particular reading of the side aisles is particularly striking as a carefully considered and beautiful effect: following the Meal – the central act of Christian Worship, the Holy Communion – one's walk back is in a field of light, brighter than when one entered. The worshipper approaches the altar in darkness, and leaves in light.

The geometry of the building is more than a conceptual framework for the liturgical act or a way to facilitate the spiritual drama of this worship: it is also a demonstration of a carefully considered set of technical solutions. Christ Church has been called one of the finest examples of acoustic finesse in American architecture. Eliel collaborated with the acoustic consultants Bolt, Beranek Newman (BBN) in the design of Christ Church Lutheran's acoustic qualities. BBN was founded by Leo Beranek and Richard Bolt (two professors at MIT), who partnered with a former student, Robert Newman, to form the company. Their first major consulting commission was the acoustic design for the United Nations General Assembly Hall in 1949,[4] but archival material from Christ Church Lutheran notes their involvement with Eliel Saarinen on Christ Church.

The geometries of the building carefully attenuate any acoustic resonance that might result in echoes or unwanted reverberation. No surfaces are truly parallel (fig. 10). The north wall of the Sanctuary, with its variegated surface, is angled slightly off the orthogonal. The east wall is also turned in, the chancel wall is curved, and the canted ceiling, both in the sanctuary, and over the side-aisles, have been carefully considered. The choir loft of the balcony edge is angled forward too. The perforated ceiling has been designed with a pattern of sound absorption panels as well as void spaces, to allow for a richer, truer auditory experience. No electronic augmentation or microphones are necessary within Christ Church; one can comfortably hear a sermon or presentation without amplification. The sensory experience is remarkable and the choral effect of the music is outstanding. One is, in a way, truly surrounded by music and by the Word. The sanctuary is a receiver both of light and of sound.

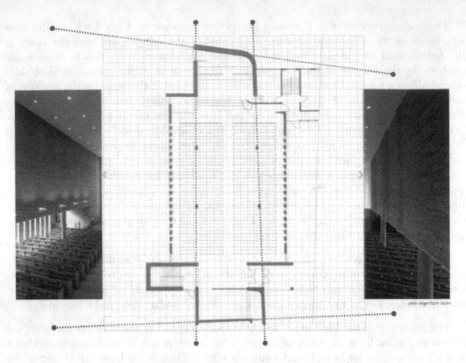

Fig. 10. Building geometry, composite image by author,
with photographs from P. Sieger, AIA and Tom Dolan

The building is never lopsided. Given its sometimes subtle and apparent shifts, it manages to achieve, in a special way, a balance and a very human scale through its asymmetries. This appropriateness of scale continues when we return to the exterior; the side aisle on the south side immediately presents a comfortable pedestrian scale to the street edge; as opposed to, for example, the massiveness of the sanctuary wall coming straight down. This creates a kind of cove, a pedestrian-friendly volume pulled from the sanctuary extending out in a gesture to the street, in recognition of the quiet residential quality of the Longfellow neighborhood. Even with a five-story campanile, the church fits very nicely in this part of town. This attention to human scale is again reiterated on the north side of the sanctuary, with the courtyard faced by the east side aisle, giving a more comfortable and humanizing scale to the garden. Eliel Saarinen continues his nuanced manipulation of scale, further breaking up the mass of the sanctuary through a simple line on the north and south walls – a single brick header extrusion – suggesting the volume of the chancel within. Instances of how the scale architecturally "speaks" about space are found on the outside of the church as well (fig. 11).

Eliel Saarinen was unlike the architectural crop working at the time. In 1949 the year the Eames House was built and Phillip Johnson's Glass House was completed, Wallace Harrison was collaborating, with, among others, Le Corbuser, Ernest Cormier and Oscar Niemeyer on the United Nations Building in New York, and Mies van der Rohe was working on the Lakeshore Drive Apartments in Chicago and the Farnsworth house in Plano. Mies and Gropius came to the US in 1937; by the time they arrived, Eliel Saarinen had been in the US for almost fifteen years.

Fig. 11. Building exterior. Image © P. Sieger, AIA and Tom Dolan, Minneapolis, Minnesota

Eliel, on the heels of his second place award in the Chicago Tribune competition in 1923, was working in the United States a few years after the first World War and during the Second. He was fully immersed in a quiet, but conscious pursuit of redefining the post-war architectural landscape. If the modern movement at the time emphasized form, structure, and the "coolly" rational, Saarinen proved with Christ Church that he was quite unlike the modernists of the time. When Frank Lloyd Wright called Eliel Saarinen, "The best of the Eclectics," Saarinen labeled him "Frank Lloyd Wrong" [Art: The Maturing Modern 1956].

Saarinen was a master of context and place, establishing an appropriateness of architectural expression for a buildings site, distinguishing him from the *Zeitgeist*. Today, none of us would argue about the importance of an architecture of context and Eliel Saarinen was ahead of his time with regard to this as well. This is reflected in the early work he did in Finland. The apartments at 17 Fabianinkatu or the projects in Katajanokka, the Railway Station and the National Museum, for example, are considered subtle and nuanced departures from the prevailing National Romanticism that preoccupied many of the architects working in Helsinki and Finland at the time. Preoccupied with this tricky question of style, his desire for his own architectural expression was expressed in the book he completed in 1948, *The Search for Form in Art and Architecture*. He believed in collaboration – as evidenced by how his office functioned – and not in the role of the architect as a heroic figure working alone. Saarinen said:

> Architecture embraces the whole form-world of man's physical accommodations, from the intimacy of his room to the comprehensive labyrinth of the large metropolis. Within this broad field of creative activities, the architect's ambition must be to develop a form language expressing the best aims of his time – and of no other time – and to cement the various features of his expressive forms into a good interrelation, and ultimately into the rhythmic coherence of the multi-formed organism of the city [Christ-Janer 1979: xvii]

This particular observation is the heart of his accomplishment at Christ Church Lutheran because the building privileges an awareness of spirit through a careful relationship between effect and cause, through a clear and deft touch with material and light – the creation ultimately, of sonorous space – both as a crafted architectural strategy and as a rich and lasting experience. The building demonstrates – and this is one of its great effects – the manner in which the masonry is held up with light, almost dissolving through the experience on the inside.

Completed at the same time as the Johnson's Glass house, Saarinen ultimately eschews the overt character of the form itself, as the structure – that physical causal reality of all architecture and in the case of Christ Church Lutheran, steel – is completely hidden through the material and surface elegance of the brick itself (figs. 12, 13). If architecture is achieved by the thoughtful resolution of opposites, then Eliel Saarinen, here at Christ Church Lutheran does it in spades. Light and the surfaces that are carefully designed to receive and hold it, its conceptual experience versus its programmatic reality, its mass versus its form, its effect versus its cause, all serve the desire of the architect to create a space of manifest spirituality. If architecture is a battle plan against gravity, then Eliel Saarinen, with the quiet and restrained poise of an architectural acrobat gives, us a space where light trumps steel and brick.

In 1948, Saarinen's second book, *Search for Form: A Fundamental Approach to Architecture* was published. 1948 was also the year that Albert Christ-Janer's *Eliel Saarinen: Finnish-American Architect and Educator* was published, "...the book that he (Eliel Saarinen) and his wife Loja both approved" [Christ-Janer 1979: xv]. Coincident with this – the publication of Christ-Janer's biography and *The Search for Form* – Eliel Saarinen would begin to design, with Saarinen, Saarinen and Associates, and the Minneapolis firm of Hills, Gilbertson and Hayes, Christ Church Lutheran.

Fig. 12. Building construction, view of side aisle and sanctuary from street.
Image © Christ Church Lutheran Archives

Fig. 13. Building construction, view of side aisle and sanctuary from campanile.
Image © Christ Church Lutheran Archives

Saarinen begins his book with the observation that the search for form, if "sincere and honest" [Saarinen 1985: v] is a simultaneous process of creation and analysis; that the creative process is the immediacy of design married to a developing system of personal, critical reflection. Saarinen's ideas of form, their genesis and evolutionary paths, logics and theories to the transitional imagination of man (the dogmatic to the mechanized to the creative mind) are all detailed here. This book is the amalgam of over a half century's experiences, reflections and musings, all directed at the principal meditation of how and why we make the things we do and essentially, of art as a total experience, rooted in the primacy of nature.

That book is not an historical overview of this thinking, nor it is a scholarly treatise on the subject. It is an attempt by Saarinen to impart a broad sense of what constitutes form-making to the young artist, this after more than quarter of a century of teaching, and more than fifty years of architectural practice, begun in 1894 in Finland. Saarinen's primer on the search for form which, as both the title and content of his book, is a combination of a reflection of his life as an architect and as an educator.

His early projects in Finland – those done first with Armas Lindgren and Hermann Geselius and later on his own – evidence the sensibility of forward thinking that has characterized much of Eliel Saarinen's work. While his contemporaries were still deeply connected to ideas rooted in National Romanticism, Art Deco and Jugendstil, Eliel Saarinen was quietly, but very consciously engaged in the search for an appropriate architecture through a deliberate dialogue with material and context. Christ-Janer references Johan Sirén's (head of the Architecture Department in Helsinki) 1955 opening address to Saarinen's memorial exhibition in Helsinki:

> The early growth of Eliel Saarinen reveals the inner relationship between personality and style. It is also convincing proof, as it always is when an artist is born, that stylistic expression is only a surface phenomenon; the impulses of the soul of the architect go deeper. The quality of expression is the constituent that decides the artistic, the ultimate value. This quality was evident during his early years [Christ-Janer 1979: 7].

Christ-Janer further notes that although the firm of Geselius Lindgren Saarinen were resistant to the embellishment of surface so characteristic to Art Nouveau, the three young architects embraced the "inventive spirit of the trend" [Christ-Janer 1979: 9].

This search for an appropriate context of architectural expression is the theme of Saarinen's life as an architect. He said:

> Most urgently something had to be done to build an art form out of the eternal fundamental principles and to bring architecture and design, in general, out of its humiliating state. In order to achieve this goal, it was necessary to free design from the its style grip and to let it develop in full freedom according to the nature and character of the time" [Christ-Janer 1979: 25].

Christ Church presents a unique opportunity to evaluate the work of two architects over a period of time, and explore their work critically and creatively in relation to the other. Even though Eero Saarinen was partnered with his father, the sanctuary for Christ Church is predominantly Eliel's hand. In 1953, in a *New York Times* article titled, "Now, Saarinen the Son" published on April 26th and written by Aline Bernsetin Loucheim, then an associate art editor, whom Eero would marry a year later, Eero Saarinen observed of his father that,

I often contributed technical solutions and plans, but only within the concept that he created. A better name for architect is 'form-giver' and until his death in 1950, when I started to create my own form, I worked within the form of my father [Loucheim 1953].

The competition for the St. Louis Arch was perhaps the inflection point for this realization. Initially awarded to Eliel Saarinen (the award notification was sent to "E. Saarinen), the correction was noted shortly after and Eliel celebrated his son's coming into his own with a grand party at Cranbrook. Designed by Eero in 1947, a year before the contract for Christ Church was issued, Eero Saarinen would not live to see the Arch built; construction was only begun in 1963 and finished five year later. The Jefferson Memorial competition coincides with Eero's burgeoning assumption of increased responsibility in his and his father's shared architectural practice. Robert Clark and Andrea Belloli write that:

> Eero Saarinen returned to Cranbrook during the summer of 1936 and entered into practice with his father. During the next three years, they produced a series of designs that marked the end of the elder Saarinen's transitional phase, established the manner of expression that characterized his last decade of practice, and served as a point of departure for Eero's independent work [Clark and Belloli 1984: 65].

Clark and Belloli also observe that Eliel's genius lay in "craft and synthesis rather than innovation" [Clark and Belloli 1984: 68], compared to Eero's characteristic ingenuity, inventiveness and his aggressive pursuit of new measures in form and technology in architecture.

When clarifying some issues for Saarinen's biographer some years after Eliel's death, Joe Lacy noted that:

> As you probably know, the last three or four years before Eliel's death were more or less transitional. Eero assumed more and more responsibility until eventually Eliel seemed content to let Eero lead the way. However, Eero had great respect for his father's ability and I doubt if Eero ever ignored Eliel's ideas.[5]

In a letter two months later, Lacy wrote that:

> The very last building that was designed by Eliel and which was built was the Christ Lutheran Church in Minneapolis That was entirely Eliel and Eero had very little to do with it. The pastor of the congregation, Rev. William Buege, is a brilliant man. He made an exhaustive search for the right architect and was finally convinced that Eliel was the right one.[6]

If Eero ever felt shadowed by the form of his father, he quickly demonstrated the power of his own architectural trajectory following his father's death, distinguished even in the way their practices operated; Eliel embodied the very description of a master in his atelier, sanguine, grand, dapper, while contrasted with Eero, rumpled and intense in his office (fig. 14), often working 18-20 hours a day, and relying on the collaborative spirit of his now famous employees: Pelli, Parker, Paulsen, Lacy, Roche, Dinkeloo, among many others. Although Eero had achieved a notable amount of fame with his early furniture projects, the Jefferson Arch competition was an important defining moment in the relationship between father and son. Their studio was "divided with a wall of blankets, pillows and sheets," so as not to reveal the developing designs of father and son to each other.[7] Indeed, Eero would significantly revise the design and detailing of the General Motors Technical Center in Warren, Michigan following his father's death – clearly demonstrating a commitment to his own ideas compared to the original scheme of his father's. Originally a shared project between father and son, the GM Technical Center became one of Eero's career-defining projects, eventually leading to many other corporate commissions.

Fig. 14. Eero Saarinen's office. Image © Leonard Parker, AIA

From his remarkable entry into the St. Louis Memorial Arch competition,[8] Eero Saarinen was one of the pioneering spirits in redefining the American Architectural landscape (fig. 15). Out of the approximately forty "25 Year Awards" given by the American Institute of Architects, Eero won five during his career. Growing up under the gaze and desk of his father, his career – although tragically cut short at the age of 51 – and contributions to the modern architectural canon cannot be overstated.

Saarinen, Saarinen and Associates continued until Eliel Saarinen's death in July 1950, shortly after the dedication of Christ Church Lutheran. That ultimately saw the full flowering of Eero Saarinen's great capacity as an architect, as well as his discipline and superhuman commitment to work, on one project committing approximately $12,000.00 towards a competition whose first prize award was $4000.00. It was noted that:

> In a single evening he will run through 170 ft. of tracing paper" and he made more than 2,000 drawings in revising his plan for the London embassy. A woman in his office, whose desk Saarinen sometimes uses late at night, inevitably knows when he has been there. Says she: "It's like slicing down through the excavations at Troy—tracing paper, tobacco,

paper, paper, matches, more paper, a cigar stub, paper, paper, paper [Art: The Maturing Modern 1956].

This work ethic would yield some of the most celebrated architectural works during the period between his father's passing and his own in 1961.

Fig. 15. The St. Louis Arch. Image © Leonard Parker, AIA

Fig. 16. The TWA International Flight Center, model. Image © Leonard Parker, AIA

Perhaps best known for his iconic designs for the Gateway Arch in St. Louis and the TWA International Flight Center at JFK Airport (fig. 16), Eero's oeuvre is impressive. He designed the John Deere Headquarters in Moline, Illinois, but died a week before construction commenced on the almost 700-acre campus. He designed the GM Technical Center in Warren Michigan, the CBS Building – the Black Rock – in New York, the IBM Plant in Rochester, the Bell Labs Complex in New Jersey, the Miller House in Columbus, Ezra Stiles College and Ingalls Rink at Yale, and Washington Dulles Airport. The echo of his father's influence can be best seen in the Kresge chapel on the MIT campus. Quiet, restrained and singularly nuanced, much of the careful quality of Christ Church is evidenced in this gem – in particular, the nuanced use of transparent and translucent views in the entry – akin to Christ Church's windows – and in the careful attention to brick as a surface. From the TWA International Flight Center to the Tulip Chair, like his father, Eero was adept at many scales.

Through all of this emerging body of work, and seemingly embodying the paradox and complications of being at once a mainstream architect, a multi-vocal designer and an innovative stylist, Eero Saarinen would return to the site of his father's last built work, and would build on his father's capacity for understanding effect and cause and the appropriateness of scale. He would design and supervise – at least the early stages of this project, because he died before this building was completed – this addition to Christ Church Lutheran (fig. 17). One can imagine that, at what was really the height of his career, when Eero Saarinen received a request to add to his father's building, the architect of the Jefferson Arch and the TWA Terminal, perhaps, straightened his tie, smoothed down his hair and honored the building and the man that ultimately transformed the direction of ecclesiastical architecture in the United States.

Fig. 17. Eero and Paulsen's addition, with the sanctuary beyond.
Image © P. Sieger, AIA and Tom Dolan, Minneapolis, Minnesota

The relationship between Eliel and Eero Saarinen is a timeless evocation of the enduring bond between father and son. In their own ways, they represented broadly different temperaments in period, style, attitude and time. Eliel was the genteel, cultured aristocrat, educator and writer, whose deft drawing hand and ideas of architecture as a synthetic craft quietly suggested a new direction for American modernism, while Eero was the intense, focused and inward-looking innovator who came to symbolize the trailblazing enthusiasm of post-war architecture in the United States. Eliel's oeuvre was identifiably consistent; it is easy to trace the formal and architectural lineage from one projects to another; Eero's body of work is vastly different in form, resulting in early scathing criticisms from the likes of Vincent Scully (cf. [Romàn 2003: 2]) who accused him of having no consistent language and said that his adaptability (now acknowledged) signified a lack of discipline!

Eero's addition to his father's work, with the significant contributions of Joseph Lacy, Glen Paulsen, Leonard Parker and Cesar Pelli,[9] is – to use a current of the theme of this writing – perhaps the most appropriate scale of effect to Christ Church Lutheran. It is modest and utilitarian, in the best senses of those words, and playful too, with a veritable Knoll and Eames showroom fronting 34th Avenue (fig. 18), but it is understated and deferential to the form of his father. Glen Paulsen, interviewed for the developing monograph on this building, noted that Saarinen's driving concern was the appropriateness of the Education Wing as a quiet response to the sanctuary. Paulsen's office worked with Eero Saarinen on this project, and Glen Paulsen[10] served as the primary designer, with Eero as the deciding voice on the project.

Fig. 18. The Education Wing's "Knoll Showroom."
Image © P. Sieger, AIA and Tom Dolan, Minneapolis, Minnesota

The addition, although programmatically different from the sanctuary, is connected to it through materiality and detailing. The great success of Eero's and Paulsen's addition lies in the apparent unity of the Education Wing with Eliel's sanctuary, but also shows an almost, I believe, tongue-in-cheek response to the work of "The Great Man," as Eliel was known, or as he was more affectionately called by his students and employees, "Papi."

Fig. 19. The sanctuary seen through an atrium window in the Education Wing. Image © P. Sieger, AIA and Tom Dolan, Minneapolis, Minnesota

Fig. 20. Christ Church Lutheran, exterior view of the sanctuary.
Image © P. Sieger, AIA and Tom Dolan, Minneapolis, Minnesota

While Eliel Saarinen's sanctuary is a profoundly inward-looking building, the Education Wing, particularly on the main floor, opens up to its context.

The floor-to-ceiling windows of the Luther Lounge (fig. 19), running the full length of the east facade of the Education Wing, open to the residential life of the street. Eero Saarinen and Paulsen turns the interior hallway that separates the classrooms into a windowed and sky-lit atrium; the main hallway that links the new entry around the small courtyard, past the church offices and into the sanctuary, is a glazed cloister. The basement is admittedly more utilitarian, but the spaces are still very social, with two lounge areas, connected to an industrial (and elegant) kitchen, linked to huge gymnasium, whose bulk and mass surpass that of the sanctuary, but is ultimately not part of the experience of the building's exterior.

While the sanctuary is a quiet volume whose experience is wholly interior and is an expression of a poetic understanding of light and liturgy, the education wing is more extroverted – in an introverted sort of way. As a tribute to his father (and in a way, as Paulsen's tribute to his friendship with Eero Saarinen), with the Education Wing, Eero has managed to reflect his own attitude about architecture in a way that honors Eliel, but also echoes his own vision for architecture. There is a unity at Christ Church Lutheran that is not only the unity of Eliel Saarinen's Cranbrook, of views and changes of scale and, as Paul Goldberger noted, of "creating a constant sense of surprise" [Goldberger 1981: 311]. Certainly the sanctuary embodies these notions, from its exterior simplicity, to its interior views and its experiential complexity. The unity at Christ Church Lutheran is of the balance of opaque and transparent, of interiority and exteriority, of the restrained craft of Eliel, and in a wonderful surprise – an innovation of sorts – of the subtle and quiet hand of Eero.

When asked what his favorite building was, Eliel Saarinen replied, "The next one",[11] anticipating that the work of the architect is never truly complete, that the search for form is unending. Christ Church Lutheran sits, in Susan Saarinen's words "quietly there," serving as a sensitive tribute to the broad and unique genius of both father and son (fig. 20). Eliel Saarinen sought specificity and appropriateness through scale, but with Christ Church Lutheran he achieved a kind of spiritual universality, and the addition by Eero Saarinen and Glen Paulsen continues that essential quality. The education wing is not flamboyant by any stretch of the imagination, but extends in a generous arc from the work of Eliel's Sanctuary, bounding the courtyard and deftly borrowing and interpreting the father's work – a curved wall, a slight reveal between stair and edge, a carefully scaled facade – while maintaining, very discreetly and informally, the authorship of the son.

Born on the same day thirty-seven years apart, August 20th, and collaborators on the same projects for some time, both Eliel and Eero Saarinen died of similar causes. They were both awarded the AIA Gold medal – Eliel Saarinen in 1947, Eero posthumously in 1962. Eero would accept the RIBA Gold Model for his father's contribution to architecture in September of 1950 just two months after the death of Eliel on July 1st of that year. Eero died in 1961. But for this text and these reflections and importantly, for the experience of the building itself, the church complex constitutes, in the buildings that embrace a courtyard, the lasting continuity of the relationship between father and son.

Notes

1. Susan Saarinen made this comment in September of 2008 at a lecture she delivered at the Minneapolis Institute of Art. Named as one of the Finlandia Foundations co-lecturers of the year (with Marc Coir), Ms. Saarinen observed – it was her first visit to the complex designed by her father and grandfather – that it was one of her favorite buildings in their oeuvre.
2. Pastor William Buege, in a letter dated November 2005 to Rolf Anderson. Mr. Anderson, a local Minnesota historian, generously shared this correspondence with me.
3. Paraphrased from Hockney's comments in the excellent documentary film, "A Day on the Grand Canal with the Emperor of China or: Surface is Illusion but so is Depth" [Haas and Hockney 1989].
4. See the BBN Technologies website at http://www.bbn.com/about/timeline/ (accessed 28 March 2010).

5. Letter from Joseph Lacy to Albert Christ-Janer, dated January 2nd, 1971, Box 2, Folder 2.1 Saarinen Family Archives.
6. Letter from Joseph Lacy to Albert Christ-Janer, dated March 18th, 1971, Box 2, Folder 2.1 Saarinen Family Archives.
7. Susan Saarinen, conversation with the author, Minneapolis 2008.
8. According to Marc Coir, Eero never gave credit to Carl Milles for suggesting that the section of the arch be triangular when Eero had requested his advice concerning his entry into the competition. Coir noted this in his lecture at the Minneapolis Institute of Art in September 2008.
9. The author has conducted several interviews with Leonard Parker, who, as the site supervisor on the addition, noted Pelli's contributions to the developing design in Glen Paulen's office.
10. It's interesting, albeit incidental, to note here that Eero recruited Paulsen personally to join the Saarinen office. Paulsen's first projects included working with Eliel Saarinen, but later transitioned to joining Eero on his work. Paulsen, notably, became the third president of Cranbrook, after Eliel Saarinen and Zoltan Sepeshy. He was interviewed in November 2008 for a developing book on Saarinen (of which this paper will form a part) in November 2009.
11. Marc Coir made this observation in his presentation with Susan Saarinen at the Minneapolis Institute of Art in September 2008.

References

Art: The Maturing Modern. 1956. *Time Magazine*, July 2, 1956.
CHRIST-JANER, Albert. 1979. *Eliel Saarinen: Finnish-American Architect and Educator*. Chicago: University of Chicago Press.
CLARK, Robert Judson, and Andrea P.A. BELLOLI. 1984. *Design in America: The Cranbrook Vision, 1925-1950*. New York: Harry N. Abrams Publishers.
HAAS, Philip and David HOCKNEY. 1989. *A Day on the Grand Canal with the Emperor of China or: Surface is Illusion but so is Depth*. Directed by Phillip Haas and written by David Hockney. Documentary film.
GOLDBERGER, Paul. 1981. Eliel and Eero Saarinen. In *Three Centuries of Notable American Architects*, Joseph J. Thorndike, Jr., ed. New York: American Heritage Publishing Company.
LOUCHEIM, Aline. 1953. Now Saarinen The Son. *New York Times*, April 26, 1953.
ROMÀN, Antonio. 2003. *Eero Saarinen: An Architecture of Multiplicity*. New York: Princeton Architectural Press, 2003.
SAARINEN, Eliel. 1948a. *The Search for Form in Art and Architecture*. Rpt. 1985, New York: Dover Publications.
————. 1948b. The Search for Form: A Fundamental Approach to Architecture. New York: Reinhold Publishing.

About the author

Ozayr Saloojee is an Assistant Professor of Architecture at the University of Minnesota's College of Design. He joined the faculty in 2005, after teaching and practicing architecture in Ottawa, Canada, where he completed his B.Arch and M.Arch degrees at Carleton University's School of Architecture. Born and raised in Johannesburg, South Africa and in Canada, Professor Saloojee's current academic areas include questions of tradition and modernity in Islamic Architecture, and research interests in contested landscapes, and the political agency of the architect in these conflict terrains. He participated with the Walker Art Center (Minneapolis) and Minneapolis Institute of Art (Minneapolis) with the hosting of the Eero Saarinen retrospective "Shaping the Future" in 2008 (as a co-organizer and invited presenter of a related symposium). He also collaborated on the exhibit "Christ Church Lutheran: Three Photographic Visions" with VJAA (Minneapolis) Site Assembly (St. Paul). The exhibit design was awarded an American Institute of Architects Honor Award in 2009. Professor Saloojee is currently completing a monograph on Eliel and Eero Saarinen's Christ Church Lutheran.

Luisa Consiglieri

Painel da Lameira Cx 8
7300-405 Portalegre
PORTUGAL
lconsiglieri@gmail.com

Victor Consiglieri

Faculty of Architecture
Technical University
Lisbon, PORTUGAL

Research

Morphocontinuity in the work of Eero Saarinen

Abstract. Continuity as the mathematical tool in the creation of architectural forms is known as morphocontinuity. In the present work, we explain how morphocontinuity appears on the work of Eero Saarinen and discuss its correspondence with its environmental (physical, social and cultural) contexts.

Keywords: Eero Saarinen, mappings, transformations

Introduction

The buildings designed by Eero Saarinen are constituted by continuous forms that have the particular characteristic of unifying the internal and external spaces. The equilibrium of the compositions is achieved by the relationship between the visual spaces and the existent movement between them.

Here we discuss Saarinen's Kresge Auditorium at MIT, the Ingalls Rink at Yale University, and the TWA Flight Center at JFK International Airport in terms of morphocontinuity. The thin shell structures are interpreted as continuous forms and these forms appear mathematically as the membranes of cells [Consiglieri and Consiglieri 2009]. These new forms supersede discrete processes such as breaking up and scattering. Rather, they are characterised by fluidity and unity. Moreover, the architectural design flows into vaster urban areas than mere buildings. This development coincides with our concept of morphocontinuity. The movement and complexity transmitted by the surfaces have a new meaning. The main achievement is then that the building is no longer a closed object but is now an open one. The symbiosis between cities and nature is achieved when no boundaries exist between private and public domains, and between interior and exterior spaces. The form can give rise to a multitude of effects which offer light and sound behaviors in a wide variety of potential directions. The inner spaces, which flow one into one another, are becoming a concern in the reconfiguration of cities. The process of production of the buildings is the result of the interaction between organization and material.

Morphocontinuity is correlated with the morphologic rhythm. It is the four-dimensional framework due to the self-organized life process of a cell [Consiglieri and Consiglieri 2006]. The mathematical cell is a 4D element reflecting cyberspaces. Actually, when we move about in a city, whether inside or outside the buildings, the act of moving brings the fourth dimension into the behavior. This new organization provides the input for the ongoing existence of cities. With the morphologic rhythm, the cell and its correspondent form provide the dynamism for the human living in the different environments.

The present study uses mathematical mappings as a means of analysis. Moreover, their graphical representation shows the visual contrasts of fullness/emptiness and brightness/darkness. Although not all of Saarinen's projects include Euclidean forms, the architectural forms have such differentiability properties that equilibrium is reached. Finally, the aesthetic value of the object is determined by the continuity of the forms and the relationship between inside and outside.

Nexus Network Journal 12 (2010) 239–247   NEXUS NETWORK JOURNAL – VOL.12, No. 2, 2010 **239**
DOI 10.1007/s00004-010-0026-4; *published online* 5 May 2010

The mathematical framework

In this section, we introduce some definitions for a fixed positive time parameter, T>0.

Definition 1. We say that a set C is a *cell* if it is defined by

$$C := \{(x, y, z) \in IR^3 : f(x, y, z; t) \le 1\}, \qquad (1)$$

with x, y, z representing (as usual) the spatial Cartesian coordinates and f denoting a continuous mapping defined in the three-dimensional Euclidean space and dependent on a time parameter $t \in [0, +\infty[$ (for details, see [Consiglieri and Consiglieri 2006]).

Fig. 1 displays four 2D-cells, namely the graphical representation of circles of radius 1:

(a) at instants of time $t = 0$ and $t = 2$ when $f(x, y; t) = x^2 + (y - t)^2$.

(b) at instants of time $t = 0$ and $t = 1$ when $f(x, y; t) = x^2 + (y - t)^2$.

(c) at instants of time $t = 0$ and $t = 1$ when $f(x, y; t) = |x|^{2-t} + |y|^{2-t}$.

(d) at instants of time $t = 0$, $t = 1$ and $t = 2$ when

$$f(x, y; t) = \left(|x|^{t^2/2 - 3t/2 + 2} + |(2t+1)y - 5t^2|^{t^2/2 - 3t/2 + 2} \right) / (1 + t(t-1)).$$

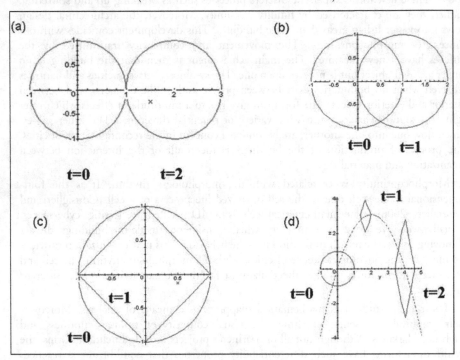

Fig. 1. (a) Euclidean rhythm: translation and juxtaposition; (b) interpenetration of congruent objects; (c) self-organized life cell; (d) movement and shape change

Definition 2. We say that a cell C has a *morphologic rhythm* if

$$\exists R > 0, \quad C(t) \subseteq B_R(0), \quad \forall t \in [0,T],$$

where $B_R(0)$ denotes an open ball centered at the origin (0,0,0) with radius R>0.

The above examples take one ball in 2D as a motif (see [Consiglieri and Consiglieri 2006] for 3D examples) under a rigid body motion where the transformation is such that the two balls act as the unique set (1) that has a free parameter denoting the time. In order to represent the juxtaposition (fig. 1a), choosing t=0 we get the initial ball and choosing t=2 it represents the translated ball. If we choose smaller instant of time, for instance t=1, we obtain an interpenetration of the congruent ball (fig. 1b).

The self-life of a cell may be clarified as follows. If we take the circle of radius one, its shape can freely change maintaining the quality of its boundary, the circumference, in separating its interior from its exterior, e.g., becoming a square (cf. fig. 1c). Moreover, the morphologic rhythm can have the following interpretation: the ball can even perform motion and shape changes simultaneously (cf. fig. 1d).

Definition 3. We define a *form* as the subpart of the boundary of the cell given by

$$F := \{(x, y, z) \in A \times IR \subseteq IR^3 : z = h(x, y; t)\},$$

where h is a continuous mapping defined in A depending on a time parameter $t \in [0,+\infty]$ (for details, see [Consiglieri and Consiglieri 2009]).

The mathematical space is the infinite set of points given in reference to the three dimensional Cartesian coordinate system. However, the architectonic space can be understood as an urban space given by the continuum 3D-space plus the time which are embedded in the 4D (space-time) coordinate system (see fig. 2).

 (a) t=2:   **(b) t=3:**

Fig. 2. The graphical representation of a form using $A = [-2,2]^2 \setminus \{(0,0)\}$, for the time parameter

$$t>0 \text{ and } h(x, y; t) = -\frac{x^4 y^4}{(x^2 + y^4)^t} : \text{(a) } t=2; \text{(b) } t=3.$$

The equilibrium should be understood as the aesthetic harmony of the visual and physical movements of the curves of the surfaces, i.e., of the properties of continuity and differentiability of the forms. At the moment, there is no a unique solution to a perfect equilibrium if the equilibrium is interpreted in terms of the dynamical point of view just described. In order to explain the present interpretation, a brief look back at the past is helpful. The Greeks aimed to obtain unifying effects, with the objective of abolishing and balancing the ascendant and descendant forces, without destroying the perception of the volumes. Using the same principles, the Roman edifices were more complex and appeared to be lighter, notwithstanding the fact that they were still massive buildings. During the Gothic period, the structures are dominated by arches and buttressing elements. These systems of equilibrium were structurally stable due to the application of proportions. With the dome of Santa Maria del Fiore cathedral in Florence by Brunelleschi [Argan 1955], the fusion between constructive technique and aesthetic emotion is managed by load methods. The movement, dynamism and complexity transmitted by the surfaces in the Baroque period were given by the successive repetitions of fragmented pieces, but plastic expression was an entity disassociated from the functionalities of the building. However, even in this era the essential characteristic of exuberance in architecture could be achieved by infinitesimal analysis [Frankl 1981: 177]. As long as the structure of a building was conditioned by the weight of the volumetric masses and the distribution of the loads, the form does not express its structures. The form is subordinated to the balance of its volumetrics. Only in the modern era does the structure become the form itself. It stands up to the plasticity. The plasticity introduced in architecture affords the possibility of realizing new configurations.

The Kresge auditorium

Extracted from the Euclidean object, the sphere, the concrete roof is the one-eighth of a sphere anchored at hidden abutments at three points [*L'architecture d'aujourd'hui* 1956; Wright 1989] (fig. 3).

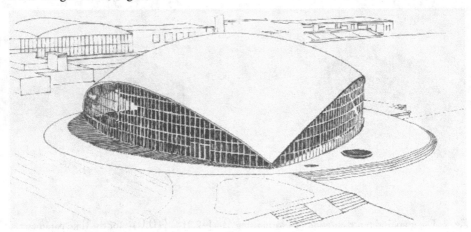

Fig. 3. Eero Saarinen, Kresge Auditorium, MIT. Drawing by Teotónio Agostinho

Under the morphocontinuity approach, the concrete shell can be understood as a form and the building as a cell (see Definitions 3 and 1, respectively). This simple form is one solution for the problem of roofing of large spaces. At the same time, it engages both the senses and the intellect. It achieves an aesthetic value, although some critics argue that the visual form is not in correlation with its function (for instance, Bruno Zevi, quoted in [*L'architecture d'aujourd'hui*: 50-51]).

Mathematically, 1/8 of a sphere can be represented as the form defined by:

$$F = \{(x, y, z) \in IR^3 : x, y \geq 0, \ z = \sqrt{R^2 - x^2 - y^2}\},$$

where the constant R>0 denotes the radius. The cell correspondent to the auditorium is not 1/8 of a sphere, but is rather the set

$$C = \{(x, y, z) \in IR^3 : x, y, z \geq 0, \ x + y + z \geq R, \ x^2 + y^2 + z^2 \leq R^2\}.$$

This cell represents the closed set constituted by the inner space surrounded by the (1/8 of the) sphere and the planar section (fig. 4) and its boundary, namely the portions of the sphere and the plane. In order to identify the set C with the auditorium in accordance with Definition 1, the set C is rotated such that the triangle becomes the ground floor, i.e., xy-plane (z=0), under the rotation over the axis located at (x,R-x,0), for all real x. The projection of the boundary of the spherical surface is no longer the triangle after rotation, but is rather the union of three arcs of ellipses.

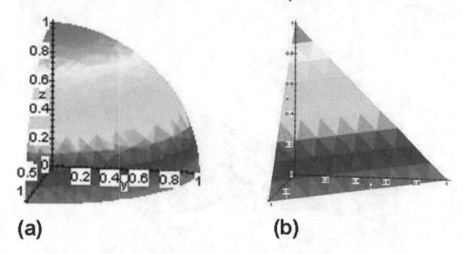

(a)   **(b)**

Fig. 4. (a) 1/8 of a sphere; (b) planar section

Moreover, the building is characterized by its volume, which is calculated as the difference between the volume integral under spherical coordinates that represents the volume of 1/8 of the ball and the one that represents the volume of the tetrahedron

$$V = \int_0^{\pi/2} \int_0^R \int_{\pi/2}^{\pi} \rho^2 \sin \varphi \ d\varphi \ d\rho \ d\theta - \int_0^R \int_0^{R-x} \int_0^{R-x-y} 1 \ dz \ dy \ dx = \frac{\pi}{6} R^3 - \frac{1}{6} R^3.$$

At a first interpretation, the spherical structure separates the inner from the outer spaces. However, the internal environment changes in accordance with different situations of the two systems: doors and windows. The sensation of space-time depends on the location of the doors, and the communication with exterior depends on the dimensions of the windows. Besides, the interconnection of the interior with the exterior is quite different if the openings are located at a lower or a higher level. At the superior level, the relationship with the exterior is panoramic. If it is located at the ground level, it creates movement toward the different directions and human activities. It reflects that the outside is always another inside, as argued by Le Corbusier [Consiglieri 1994: 95].

Here, the transparency of the glass walls dematerializes the boundary between the interior and the exterior spaces, dissolving the traditional dichotomy between inside and outside.

The Ingalls rink

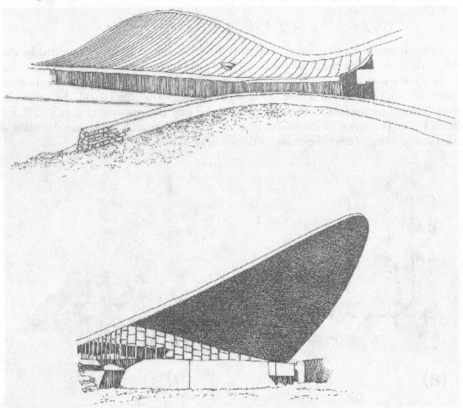

Fig. 5. Eero Saarinen, Ingalls Rink, Yale University. Drawing by Teotónio Agostinho

Given the couple composed of the rectangular rink and the seats that surround it, which forms the basis for the architectural object, the principal facade is surprisingly constituted by an awning with the boundary edge measuring 73.0 m (fig. 5). The main purpose of this shell is to funnel the audience from outside into inside through the mediating point, which is the entrance. This entrance, which faces onto and simultaneously ends at the lateral walls, flows into the walls of the parking and the external arrangements. The axis of symmetry of the suspended vault coincides with the axis of the rink and the corresponding lateral seats, establishing the unity of the inside with the outside. Thus we can say that the building is constituted by two twin cells, each of which is highlighted by a form under the continuous mapping:

$$h : A = [-1,1] \times [-1,1] \rightarrow IR$$

defined as:

$$h(x, y) = \sin(a\,x)\cos(b\,y)$$

for some constants $a, b \in IR$ (compare to the FresH2O pavilion, Netherlands, 1998, by Nox team and Kas Oosterhuis [Consiglieri and Consiglieri 2009]). The intertwining of functional spaces and facades recreates a cyclical and shared feeling around the building.

The TWA Flight Center

The roof of the TWA terminal is constituted by four free-flowing shell structures suggesting flight and having a thickness that varies from 20 cm to 1.10 m (fig. 6). This variation reveals the development of new structures and formal systems. The flying shells are continuous forms of the kind of *non-poids* in architecture theorised by modernist European architects [Robichon 1965]. The various interior levels are linked by stairs and sculptural galleries. This fluid space transforms our visual perception. Indeed, the created cells are provided by an emotional movement because the images change under our visual perception. This movement can be mathematically interpreted by the time parameter in eq. (1). The organization is commonly compared to the cyclical nature of the Möbius strip folding back onto itself. The application of topological deformation of a surface can lead to the intersection of external and internal planes in a continuous morphological change and insert differential fields of space and time into an otherwise static structure [Emmer 2004: 65].

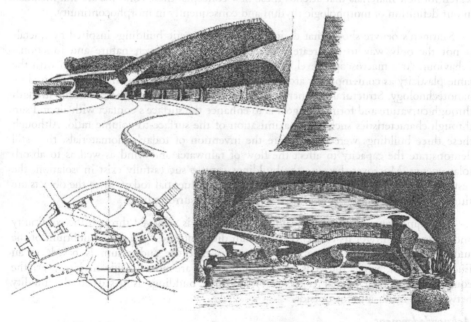

Fig. 6. Eero Saarinen, TWA Flight Center, JFK International Airport.
Drawing by Teotónio Agostinho

In addition to the glass in this building acting to dissolve the boundary between interior and exterior, as it did in the other two buildings considered earlier, here the different sized surface units are distributed such that the modulation of the interior and exterior spaces becomes irrelevant. Moreover, they organize the internal environment and maximize the penetration of sunlight. The final object is constituted not only by different forms but also by different cells, which include the interior of such forms.

Conclusions

In order to interpret the chosen works conveniently, we consider two types of interpretations for the structure:

1. the structure is understood in terms of its physical characteristics and mathematical methods. In this point of view, the structure is defined by the characteristics of the materials and the loads;

2. the structure loses its importance in relation to the final object and is driven by mathematical precision and clarity.

The works studied here have roofs of different forms. Although they are all limited by the use of concrete, the three buildings show that the dynamic search is possible. We can conclude that the study of eurhythmy is based on the concepts of continuity and differentiability of the object itself and continuous movement between interior/exterior.

As far as contemporary interpretations of architectural space (namely utopian space [Roisecco 1970: 52-53, 364-445], existential space [Norberg-Schulz 1977: 430-433], emergent, biomorphic or organic spaces [*Architectural Design* 2008]) attempt to imitate the biological behaviour of self-organization, along with the simultaneous, ongoing search for new materials that attends these new concepts, these concepts are still included in our definition of morphologic rhythm and consequently in morphocontinuity.

Saarinen's oeuvre shows that designing morphogenetic buildings inspired by fractals is not the only way to be create a perfect symbiosis between nature and functional behaviour. At a macroscopic level, the concrete structures seem to be modeled with the same plasticity as contemporary architectural objects using the new materials inspired by nanotechnology. Structurally, concrete is relatively economical, in that it affords strength through curvature and form. The goal is to enhance the surface's contact with air and sun through characteristics such as a maximisation of the surface-to-volume ratio. Although these three buildings were built before the invention of today's biomaterials, they still demonstrate the capacity to affect the flow of rainwater and wind as well as to absorb solar energy. Moreover, because the buildings cannot successfully exist in isolation, the aesthetic value of the objects discussed here is still influential today, since the objects are situated in the dynamic centre of even more dynamic surroundings.

The equilibrium of the composition of the object is achieved through the continuity and differentiability of its elements. The main difference between the equilibrium of the auditorium and that of the other buildings is that the auditorium is based on an Euclidean form, while the cells of other buildings are non-Euclidean. However, the equilibrium of their spatial elements of all three buildings is based on the fluidity between the inside and the outside.

Acknowledgement

The authors gratefully thank Teotónio Agostinho for figs. 4-6.

References

L'architecture d'aujourd'hui. 1956. Auditorium du M.I.T., Cambridge, U.S.A., Eero Saarinen et Associés, Architectes. L'architecture d'aujourd'hui 64, 27 (Mars 1956): 50-53.

Architectural Design. 2008. Neoplasmatic design. Architectural Design 78, 6.

ARGAN, G.C. 1955. Brunelleschi. Arnoldo Mondadori Editore (BMM 415), Veronesi.

CONSIGLIERI, L. and CONSIGLIERI, V. 2006. Structure of phenomenological forms: morphologic rhythm, Nexus Network Journal 8, 2 (October 2006): 101-110.

———. 2009. Continuity versus Discretization. Nexus Network Journal 11, 2 (2009): 151-162.

CONSIGLIERI, V. 1994. A morfologia da arquitectura 1920-1970, vol. 2. Ed. Estampa, Lisbon.

EMMER, M. 2004. Mathland: from flatland to hypersurfaces. Birkhäuser, Basel.

FRANKL, P. 1981. Principios fundamentales de la Historia de la Arquitectura. El desarrollo de la arquitectura europea: 1420-1900. Ed. Gustavo Gili, S.A., Barcelona 1981. Translated by Die Entwicklungsphasen der Neueren Baukenst. Ed. B.G. Teubner, Stuttgart 1914.

NORBERG-SCHULZ, C. 1977. La signification dans l'architecture occidentale. Pierre Mardaga editeur, Bruxelles. Translated by Significato nell'architettura occidentale. Electra Editrice, Milano 1974.

ROBICHON, R. 1965. Vers le non-poids en architecture. In Structures nouvelles en architecture. Conservatoire National des Arts et Métiers, Paris (Avril 1965), 10-11.

ROISECCO, G. 1970. Spazio, evoluzione del concetto in architettura. Mario Bulzoni Editore, Roma.

WRIGHT, S.H. 1989. Sourcebook of Contemporary North American Architecture: From Postwar to Postmodern, Van Nostrand Reinhold, New York.

About the authors

Luisa Consiglieri received her B.S., M.S. and Ph.D. degrees in mathematics from University of Lisbon in 1988, 1992 and 2000, respectively. She taught at the Faculty of Sciences of University of Lisbon from 1987 until her retirement in April 2009. Her research interests span a variety of areas in differential equations such as existence, uniqueness and regularity of solutions, fluid mechanics, heat transfer, electromagnetism, biomechanics, and applications to biomedical problems. At the present her principal interest is interdisciplinarianism.

Victor Consiglieri received a B.S. degree in architecture from Escola Superior das Belas-Artes de Lisboa (ESBAL) in 1956, and a Ph.D. degree in morphology of architecture from Faculty of Architecture of Technical University of Lisbon in 1993. He was in Paris 1964-65 on a scholarship from Centre Scientifique et Technique du Bâtiment (CSTB). He was in Camâra Municipal de Lisboa 1956-62, Ministério do Ultramar 1962-66, Caixa da Previdência 1966-76, Faculty of Architecture of Technical University of Lisbon 1976-97 and as invited professor in University of Évora 2004-05. He realized many projects for public buildings such as kindergartens, elementary schools, institutions for youth, and centres for the elderly. He was member of Associação dos Arquitectos and of Ordem dos Arquitectos 1956-2004. His current interest is the contemporary aesthetic.

Tyler Sprague

University of Washington
College of Built Environments
Seattle, WA 98103 USA
tyler2@u.washington.edu

Keywords: Eero Saarinen,
Eduardo Catalano, Matthew
Nowicki, modern architecture,
hyperbolic paraboloids, saddle
shapes

Research

Eero Saarinen, Eduardo Catalano and the Influence of Matthew Nowicki: A Challenge to Form and Function

Abstract. Matthew Nowicki befriended Eero Saarinen at the Cranbrook Academy and was succeeded as Chair of the School of Design at North Carolina College of Design by Eduardo Catalano. Nowicki's influence is evident in subsequent work of these two architects. Themes of function, structure and humanism resonated differently in each. All three of these interconnected individuals were engaged in the same intellectual milieu, each manifesting his own architecture in a unique yet contextual way. Taken as a whole, their endeavors stand as evidence of the shifting understanding of what modern architecture was about.

Introduction

In the years following World War II, modern architecture was a facing a cross-roads.[1] Many architects were skeptical of the ability of the dominant "International Style"[2] to respond to social, humanistic demands – a sentiment magnified by the loss of life during the war – and were searching for a new direction forward. The self-justified rational architecture did not engage the richness of human experience. Multiple calls for "expression" and "monumentality" in architecture revealed a need for more human engagement than the International Style was providing [Barr 1948].

Thus post-war modernist architects had to make a choice: maintain the rigid obedience to "functionalist" architecture, or search for a new means of expression. They were consciously engaged in re-thinking what architecture should be, interested in recasting modern architecture to suit an altered social landscape; they wanted to make architecture more appealing, but didn't want to abandon industrial efficiency. They sought to re-place architecture in the minds of citizens, but felt the "immaturity of modern architecture", and the need to "grow up".[3] Modernism had not failed, but it needed revision.

One architect who engaged in this discussion and would prove influential in altering its course was Matthew Nowicki. His untimely death in an airplane crash in the fall of 1950 cut his budding career short, but not before he had written a few influential articles and designed one seminal work, the Livestock Pavilion in Raleigh, North Carolina. Nowicki's work suggested a new direction for architecture in the post-war world – ideas that would fatefully remain unrealized by him, but would be picked up and extended other like-minded architects.

During his life, he befriended the architect Eero Saarinen at the Cranbrook Academy and was succeeded as Chair of the School of Design at North Carolina College of Design by Eduardo Catalano.[4] Through the subsequent work of these two men, Nowicki's influence – his personality, his inventiveness, his ideas about architecture – can be seen. Themes of function, structure and humanism resonated differently in each man, striking different chords in their work. Together, these three interconnected individuals, with great similarities and vast differences, were engaged in the same intellectual milieu, each manifesting his own architecture in a unique yet contextual way. Their combined endeavors, physical and intellectual, stand as evidence of the shifting understanding of what modern architecture was about.

Matthew Nowicki

Matthew Nowicki was born in 1910, into a Polish noble family, the son of a politically active Consul to the Polish State. He traveled extensively as a child, spending much time in Chicago, and learning English before enrolling in architectural studies at the Warsaw Polytechnic in 1928. Here he quickly demonstrated an incredible drawing ability, a talent that would provide his most lasting legacy in his striking sketches that remain. Instruction at the Polytechnic also strove to "teach him to see things as structures. To this end a drawing was built, with skeletons of structural lines exposed" [Mumford 1954a: 142]. Drawing for Nowicki was not just symbolic representation, but an intellectual process of architectural and structural synthesis. Engineering and geometry would always go hand-in-hand with his formal, architectural explorations. Nowicki also met his wife, Sasha, while in school; a fellow architecture student, she was by all measures his equal in drawing and design.

Inspired by Le Corbusier and Wright, Nowicki began his own professional practice in Warsaw after graduation, also accepting a position as associate professor at the Polytechnic. He built a number of churches, homes and sports arenas in Poland before the German invasion in September 1939. The darkness of the war descended on Warsaw as the mass destruction increased, and Nowicki and his family were forced to escape to the distant mountain regions. After the war, Nowicki was involved in the planning to rebuild Warsaw, but when the Polish government was taken over by Communist powers, he decided to come to the United States [Brook 2005: 37]. He served as Cultural Attache to the Polish Consulate in Chicago, was a visiting critic at the Pratt Institute, and served a formative role as the Polish representative to the committee for the United Nations Building.[5]

The wartime experience drastically effected Nowicki's outlook on life, and revised his approach toward architecture. After the war he wrote:

> The study of the well-being of contemporary man, which has been introduced into the language of architecture, continues to be the inspiration for our work but this time the quality is differently analyzed. It is no longer 'the machine to live in' that stirs our imagination. It is the eternal feeling of a shelter to which we subordinate our creative ideas [Mumford 1954: 148].

Nowicki invokes a more humanist approach to architecture, one that is sensitive to emotional feelings as well as function.

His technical education at the Warsaw Polytechnic also rooted his architectural theories in structural realities. Advancing technologies, as they changed over time, were a crucial part of creating effective architecture. He stated:

> ...we now rely in our expression of the potentialities of materials and structures. This interest in structure and material may find within the building medium decorative qualities of ornament that are much too involved for the purist of yesterday. The symbolic meaning of a support

has been rediscovered, and a steel column is frankly used as a symbol of structure, even when it is not part of the structure itself [Nowicki 1951: 279].

For Nowicki, structure had become expression and he embraced the new potential that emerging technologies suggest.[6] A new type of expression was emerging from an awareness of material and structure, one that was not rooted in a singular, formal prescription but rather encouraged multiple investigations. He states, "Art may be one, but it has a thousand aspects. We must face the dangers of the crystalizing style... trying to enrich its scope by opening new roads for investigation and future refinement" [Nowicki 1951: 279].

In an article titled "Origins and Trends in Modern Architecture," Nowicki clearly stated what many architects had begun to feel regarding functionalist architecture:

> In the growing maturity and self-consciousness of our century, we can not avoid the recognition ... that the overwhelming majority of modern design form follows *form* and not *function*. And even when a form results from a functional analysis, this analysis follows a pattern that leads to a discover of the same function, whether in a factory or a museum [Nowicki 1951: 273].

Striking to the heart of the proclaimed objectivity of pre-war modern architecture, Nowicki's statements resonated with many architects of his generation. From Paul Rudolph [1986: 153] to Colin Rowe [1976: 130], Nowicki's statements provided a springboard for a new line of architectural thinking, causing, in 1962, the critic Allan Temko to call Nowicki the "spokesman for young Modernists" [Temko 1963: 43]. For the emerging generation of architects, function influenced but did not dictate form. For Nowicki, Catalano and Saarinen, post-war architecture would engage the mutual dependence of function and form.

Nowicki's theories can be seen in his sketches for the State Fair Livestock Judging Pavilion (later the Dorton Arena) in Raleigh, North Carolina. His death during the design process left architect William Henley Deitrick, working with engineers Severud-Elsted-Kreuger, to complete the project. Despite the intent to do everything "as Matthew would have wanted it" [Parabolic Pavilion 1952: 137], substantial design changes due to budget and construction issues altered the building to the point where some questioned if Nowicki would have been pleased with it (North Carolina Dean Henry Kamphoefner quoted in [A radical settles down in Raleigh, NC 1980]. But Nowicki's intent is evidenced through his sketches, and is indicated in the built work.

The sketches (and the building itself; fig. 1) are structurally bold, consisting of two intersecting parabolic arches. The sweeping, mathematically-driven forms are angled to the horizon, and on this account, trace the plan of the arena in the area between them. The roof consists of draped cables strung between the two arches. The cables' dependence on the geometry of the arches, coupled with their own catenary behavior, creates a warped roof surface displaying a curvature in both lateral and longitudinal directions. This is clearly a geometrical investigation of shapes in space, a mixing of elevation and plan, but it is also underlined by a structural logic.

The two parabolic arches were made of concrete, and act in pure compression, with the roof cables hung in pure tension. The force-imbalance of the canted arches was to be ideally counteracted by the thrust of the roof cables, freeing up the curtain wall below to provide only stability. In the final building, construction methods had not advanced enough to support these intentions, and the exterior curtain wall became load bearing. These structural innovations have been widely cited by people such as Frei Otto [1954] as innovative and suggesting a new field of architectural form. The Livestock Pavilion displayed a material logic employed through the mathematics of complex geometry.

Fig. 1. Matthew Nowicki's State Fair Livestock Judging Pavilion (later the Dorton Arena) in Raleigh, North Carolina. Photograph by Yoshito Isono, reproduced courtesy of Nicholas Janberg and Structurae

These structural forms also provided an expression of the function of the building and shaped the space within. As a "single great room" the livestock show floor and surrounding grandstands mirror the structure above, providing axiality and a center of focus (Paul Rudolph quoted in [The great Livestock Pavilion complete 1954]). *Architectural Forum* stated:

> Nowicki was seeking first of all not for a unique structure but for a unique space. The remarkable warping of the space upward, the exact reverse of a dome, would guarantee maximum daylight admitted from the two sides to the central arena. This labile kind of curvature of enclosed space marks a new epoch in architecture [Parabolic Pavilion 1952].

The innovative, three-dimensional aspects of the Pavilion indicated a different means of enclosing human experience, a new spatial relationship within architecture. The roof provided shelter and also engaged the questioning mind.

Nowicki's theories of humanist expression through structure are in play here. He is demonstrating a navigated position between the human experience, structural rationality and the functional demands of the building. Although not without shortcomings,[7] in this single work he has embodied many different strands of the architectural discourse of the time. Paul Rudolph described a sublime experience, and stated that this new space "helped man forget something of his troubles." He also claimed that it satisfied "our need for more expressiveness to emphasize our places of worship, meeting places of governing bodies, and centers of recreation" [The Great Livestock Pavilion Complete 1954: 132]. It embodied a "basic soundness and high spirited boldness" that indicated a new direction for modern architecture [The Great Livestock Pavilion Complete 1954: 134].

Eduardo Catalano

After Nowicki's death, his position as chair of the North Carolina State College, School of Design was soon filled by Dean Henry Kamphoefner, with Eduardo Catalano. Catalano, an Argentinian-born architect who trained at both the University of Pennsylvania and under Walter Gropius at the Harvard Graduate School of Design, was keenly interested in advanced geometrical forms in architecture. Prior to coming to North Carolina, he had co-written a book on the mathematics of geometrical forms and the use of perspective [Crivelli, Nery and Catalano 1940]. His un-built auditorium project in Buenos Aires used a thin shell structure to enhance acoustics, and he entered several competitions for pre-fabricated housing solutions, utilizing a modular approach to building [Arts and Architecture's Second Annual Competition 1945].

But it all appears to come together for Catalano once he arrives in Raleigh. Along with the continuing construction of the Livestock Pavilion, Nowicki's presence was still felt through the curriculum and education system he had initiated. His pedagogical approach to teaching architecture emphasized a merge of the technical requirements with a humanistic awareness. Influenced by Le Corbusier's "Modulor", Nowicki emphasized designing around "Man" as the "unchanging module of of scale and proportion" and the role of technology and structure as a means to satisfy changing human demands [Mumford 1954b]. This legacy is digested and synthesized by Catalano, filtered through his own experiences and disposition, in a unique way.

Catalano's work focused on geometrically advanced forms, investigating new means of spanning space. He worked primarily with hyperbolic paraboloids, investigating geometrical surfaces with curvature in two longitudinal directions (e.g., saddle shapes) – forms like Nowicki's Livestock Pavilion roof. But Catalano extended this form, coming up with a modular system to create these shapes using individual, linear elements. Drawing on his experience with advanced geometry, Catalano developed a system where new shapes could be generated by simply modifying parameters of the design. His project *Structures of Warped Surfaces* explored various combinations of hyperbolic paraboloid forms with a variety of supports, searching for new ways to provide an over-head surface [Catalano Gubitosi Izzo 1978: 55-70]. In 1953, three years after Nowicki's death, Catalano was quoted stating:

> Following the research begun by Nowicki, our work at North Carolina has gone far beyond the Raleigh Pavilion in the study of both space structures and repetitive spatial structural systems ... these have led to interesting structures (comment by Eduardo Catalano, in [Is this Tomorrow's Structure? 1953: 160]).

Catalano also utilized war-related technologies of aluminum and plywood in experiments with these forms, stressing innovation tied to industrial production.

His work at North Carolina State College culminated in the construction of his own house in the woods outside Raleigh. Known simply as the "Catalano House" (1953-55, fig. 2) it consisted of a single hyperbolic paraboloid, made up of three layers of laminated timber. Spanning 90 feet between supports with a total thickness just over two inches, this house was celebrated as "a structure that is all skin and no bones," reflecting "the most advanced engineering know-how" of the time "[Why are People Talking about this House? 1955]. The functions of house take place then beneath the shell, with full height glass curtain walls defining the interior space.

Fig. 2. Eduardo Catalano's Catalano House

But Catalano's house was recognized as an experiment, an attempt to utilize advancing "skin technology in architecture" "[Why are People Talking about this House? 1955]. Given his approach to design, he was faced with the awkward challenge of fitting an architectural function to the forms he had developed through mathematics. The majority of his "warped surface" sketches contain no indicator of relation to the human scale, nor to which functions or building types they would be most suited. Nowicki's work clearly influenced his explorations, but Catalano has placed new emphasis on geometrical "purity" and advancement in structural engineering without direct ties to a specific architectural program [Catalano 2009]. His "warped plane roof" became famous with both "avant-garde aestheticians and building technicians " "[A New Way to Span Space 1955]. It is a shape like a potato chip but is also mathematical, geometrical, and analytical: a universal shape for covering.

Eero Saarinen

The parallel work of Eero Saarinen reveals a different aspect of Nowicki's influence. Saarinen and Nowicki met at a symposium in February 1948, and in the summer of 1949, Nowicki was appointed Visiting Professor at the Cranbrook Academy.[8] United by a propensity for drawing, the two were fast friends and shared a fruitful summer designing the campus plan for Brandeis University [Merkel 2005: 105]. Their sketches show heavily sculptural forms, with undulating walls and domed spaces, which, though not without precedent, mark a formalistic departure from Saarinen's previous work, like the GM Technical Center. With Nowicki by his side, Saarinen is able to merge his sculptural interests developed in furniture with the architectural humanism of Nowicki. Though their plan for Brandeis went unrealized, their sketches reveal many hints of Saarinen's future work.

Saarinen would later acknowledge the influence of their brief time together, which lasted only a few months. In a letter, he declared Matthew Nowicki to be the third most significant influence on him after his father, Eliel, and his life-long collaborator Charles Eames [Saarinen Pelkonen and Albrecht 2006: 332]. In Nowicki's obituary in *Architectural Forum*, Saarinen stated: "If time had allowed his genius to spread its wings in full, this poet-philosopher of form would have influenced the whole course of architecture as profoundly as he inspired his friends" [Mumford 1950]. Through these statements, we can see that Nowicki had a profoundly different effect on Saarinen than on Catalano.

After his experience with Nowicki, Saarinen's work is defined by his formal expression, in such work as the MIT Chapel and Kresge Auditorium, and the TWA Flight Center and Dulles Airport Terminals [Serraino Saarinen and Gössel 2005]. The structural logic of his architecture is often complicated by his attempts to create a meaningful form, one that communicates as well as functions [Pelkonen 2006]. Inspired by seeing Nowicki as a poet-philosopher of form, Saarinen was not interested in pure structure or engineering, like Catalano, but in linking architectural form to context and broader, humanist theories. Geometry alone was not enough; it needed energizing if it "was to serve the spatial-structural-spiritual totality" that he wished to express [Temko 1962: 43]. This is architecture that engages Nowicki's ideas of the multiplicity of art.

Fig. 3. Eero Saarinen's David S. Ingalls Ice Rink at Yale University, New Haven, CT. Photograph by Yoshito Isono, reproduced courtesy of Nicholas Janberg and Structurae

Nowicki's influence is most clearly seen at Saarinen's Ingalls Ice Rink (fig. 3). With a program similar to the Livestock Pavilion, Saarinen gives expression to the Ice Rink as a "single room" with a central rink surrounded by grandstand seating. A single, long concrete arch swoops over the long axis of the rink, with catenary cables running perpendicular to either side, forming the roof surface. But in an artistic and functional move, Saarinen does not terminate the main arch at its support, but reverses its curvature and cantilevers an additional portion to serve as an awning over the building entrance. This extended curve gives the building an undulating quality, making it a graceful active presence on the north side of the Yale University campus [Yale's Hockey Rink 1958]. Exposed on the interior, the smooth concrete arch contrasts with the wood plank finish of the roof, coming to its peak over the hockey rink below. Commonly called the "dinosaur" or likened to a Viking ship [Yale's Viking Vessel 1958], the building has "personality" and engages people spatially and intellectually.

Not only did Saarinen use a structural system similar to that of the Livestock Pavilion (as well as hiring the same engineer) but he captured a similar dynamic movement that can be seen in Nowicki's sketches. His structure dominates the form, serving the purpose of providing enclosure and supporting function, but clearly standing as a unique, artistic creation. The Ice Rink was criticized as not "sensible" because it spanned the arch over the long dimension (rather than the short axis) but it highlights the fact that Saarinen was not interested in a prescribed, "logical" approach to structure; the central spine served more than just a structural purpose.[9] He was not interested in the geometrical warping of overhead space as an isolated experiment, but in its inclusion, modification and distortion in the overall architectural experience.

Conclusion

The work of Eduardo Catalano and Eero Saarinen reveals different themes in post-war architecture. Ranging from systematic, geometrical form-finding to expressive, artistic architecture, these architects display different directions following their experience with Matthew Nowicki. The sketches, writings, and built work of one man were simultaneously described as the investigation of spatial structures, and the work of a poet-philosopher, spawning different yet related directions in post-war architecture. This study indicates the nature of modern architectural discourse - as the action of neither isolated, autonomous individuals nor a unified group. Nowicki's influence does not at all indicate a lack of originality or innovation on the part of either architect, but serves to show how architectural discovery is often contextual and interrelated to broader discourses. By discussing the milieu of theories and shapes, functions and forms, this comparative study reveals the changing landscape of ideas about how to design modern architecture in the post-war period.

Lewis Mumford's comments following Nowicki's death in 1950 appear prophetic yet misleading. A friend and proponent, he stated that Nowicki

> bore within him the seed of a new age. In his designs, spontaneity and discipline, power and love, form and function, mechanical structure and symbol, were united. What he left undone through his death must now call forth the creative efforts of a whole generation [Mumford 1950: 201, 206-207].

Not simply completing Nowicki's work, Catalano and Saarinen forged their own ways forward, inevitably leaning on their own experiences and influences, on the paths to creating their own unique architecture.

Notes

1. For a much longer discussion of the post-war architectural scene, see [Goldhagen and Legault 2000].
2. The title of a 1932 Exhibition at the Museum of Modern Art, followed by the publication [Hitchcock and Johnson 1932].
3. Lewis Mumford used the terms "immaturity" and "grow-up" during the 1948 symposium "What is Happening to Modern Architecture" [Barr 1948].
4. Nowicki was returning from a trip to Chandigarh, India where he was designing a new capital complex with Albert Meyer. This project would be later taken up by Le Corbusier.
5. Nowicki is credited, along with Le Corbusier and W. K. Harrison, as having a significant impact on the UN design. See [UN General Assembly 1952: 141].
6. These ideas are very similar to Louis Kahn's claims in his article "Monumentality" [1944].
7. Problems with the acoustics and waterproofing of the Pavilion proved difficult to solve.

8. They were both invited to the "What Is Happening to Modern Architecture" symposium at the Museum of Modern Art, organized by Alfred H. Barr and Henry-Russell Hitchcock. Cranbrook position cited in [Nowicki and Schafer 1973: xi].
9. Quote from Robert Venturi in "Appreciations by Former Collaborators Panel Discussion" in [Saarinen, Pelkonen, Albrecht 2006: 361].

References

A New Way to Span Space. 1955. *Architectural Forum* **103** (November 1955): 170-177.
A radical settles down in Raleigh, NC. 1980. *AIA Journal* **69**, 11 (September 1980): 54-61.
Arts and Architecture's Second Annual Competition for the Design of a Small House. 1945. *Arts and Architecture* **62** (Feb. 1945): 28-41.
BARR, Alfred H. 1948. What Is Happening to Modern Architecture?: A Symposium at the Museum of Modern Art. *MOMA Bulletin* XV, 3. New York: Museum of Modern Art.
BROOK, David Louis Sterrett. 2005. Henry Leveke Kamphoefner, the Modernist: Dean of the North Carolina State University School of Design, 1948-1972. Master's Thesis, Norrth Carolina State University. http://www.lib.ncsu.edu/theses/available/etd-07252005-164332/unrestricted/etd.pdf.
CATALANO, Eduardo, Camillo GUBITOSI, and Alberto IZZO. 1978. *Eduardo Catalano: buildings and projects.* Rome: Officina.
CATALANO, Eduardo. 2009. Interview by the author. 9 December 2009.
CRIVELLI, Oscar F., Rene NERY, and Eduardo Fernando CATALANO. 1940. *Teoria de las sombras y trazados de perspectiva.* Buenos Aires: Francisco A. Colombo.
GOLDHAGEN, Sarah Williams and Réjean LEGAULT. 2000. *Anxious Modernisms: Experimentation in Postwar Architectural Culture.* Montréal: Canadian Centre for Architecture.
HITCHCOCK, Henry-Russell, and Philip JOHNSON. 1932. *The International Style: Architecture since 1922.* New York: W. W. Norton & Company. Rpt. 1995, New York: Norton.
KAHN, Louis. 1944. Monumentality. Pp. 570-579 in The New Architecture and City Planning, Paul Zucker, ed. New York: New York Philosophical Library.
MERKEL, Jayne. 2005. *Eero Saarinen.* London: Phaidon.
Is this Tomorrow's Structure? *Architectural Forum* **90** (February 1953): 150-160.
MUMFORD, Lewis. 1950. From the legacy of Matthew Nowicki. *Architectural Forum* October 1950: 200-201.
———. 1954a. The Life, Teaching and Architecture of Matthew Nowicki. *Architectural Record* **115** (June 1954): 129-139.
———. 1954b. Matthew Nowicki as an Educator. *Architectural Record* **116** (July 1954): 128-135.
NOWICKI, Matthew. 1951. Origins and Trends in Modern Architecture. *Magazine of Art* **44** (November 1951): 273-79.
NOWICKI, Matthew, and Bruce Harold SCHAFER. 1973. *The Writings and Sketches of Matthew Nowicki.* Charlottesville: University Press of Virginia.
OTTO, Frei. 1954. *Das hängende Dach* (The Hanging Roof). Berlin: Bauwelt Verlag der Ullstein.
Parabolic Pavilion. 1952. *Architectural Forum* **97** (October 1952): 134-139.
PELKONEN, Eeva-Liisa 2006. The Search for Communicative Form. Pp. 83-07 in *Eero Saarinen: Shaping the Future.* Eero Saarinen, Eeva-Liisa Pelkonen, and Donald Albrecht. New Haven: Yale University Press.
ROWE, Colin. 1976. *Mathematics of An Ideal Villa and Other Essays.* Cambridge MA: MIT Press.
RUDOLPH, Paul. 1956. The Six Determinates of Architectural Form. *Architectural Record* **120** (October 1956): 183.
SAARINEN, Eero, Eeva-Liisa PELKONEN, and Donald ALBRECHT. 2006. *Eero Saarinen: Shaping the Future.* New Haven: Yale University Press.
SERRAINO, Pierluigi, Eero SAARINEN, and Peter GÖSSEL. 2005. *Eero Saarinen, 1910-1961: A Structural Expressionist.* Köln: Taschen.
TEMKO, Allan. 1962. Eero Saarinen. Makers of contemporary architecture. New York: G. Braziller.
The Great Livestock Pavilion Complete. 1954. *Architectural Forum* **100** (April 1954): 130-134.

UN General Assembly. 1952. *Architectural Forum* **97** (October 1952): 140-149.
Yale's Hockey Rink. 1958. *Architectural Record* **124** (October 1958): 151-158.
Yale's Viking Vessel. 1958. *Architectural Forum* **109** (December 1958): 106-111.
Why are People Talking about this House? 1955. *House and Home*, August 1955: 94-100.

About the author

Tyler Sprague is a doctoral student at the University of Washington, Seattle. He has a background in structural engineering and continues to study architecture and engineering in the post-war period. He would like to thank Meredith Clausen and Alex Anderson for their guidance, and acknowledge support from the University of Washington Victoria N. Reed Endowed Student Support Fund.

Rachel Fletcher

113 Division St.
Great Barrington, MA
01230 USA
rfletch@bcn.net

Research

Eero Saarinen's North Christian Church in Columbus, Indiana

Keywords: Eero Saarinen, North
Christian Church, descriptive
geometry

Abstract. Eero Saarinen's North Christian Church, an important contribution to post-war liturgical church architecture, serves a community of Disciples of Christ in Columbus, Indiana. Early design sketches illustrate elementary geometric shapes and symbols – triangle, square, cross, hexagon, and octagon – whose proportions appear in the plan and section of the completed structure.

Introduction

> *...they should feel they are all in unity and harmony in a special and appropriate spiritual atmosphere.*
>
> Eero Saarinen, on North Christian Church (1960) [Saarinen 1962: 88]

An important contribution to postwar liturgical church architecture, Eero Saarinen's North Christian Church serves a northern residential suburb of Columbus, Indiana. Designed between 1959 and 1961 and completed in 1964, the central plan church accommodates greater participation and more intimate and democratic congregational involvement.

Saarinen desired a simple structure that would support the liturgical needs of its members and "clearly and logically express the form and character of the church" (on North Christian Church (1960) [Saarinen 1962: 88]). Its open structure, facilitated by new advances in steel construction, dissolves the traditional distinction between officiate and congregant, while recognizing communion and baptism as important liturgical sacraments.[1]

Exterior

Saarinen designed the church at the request of J. Irwin Miller, following the architect's design for Miller's private residence (1953-57).[2] In 1955, forty-three members of First Christian Church broke away to organize a more liberal institution affiliated with the Disciples of Christ. In 1958, with Miller's assistance, the group purchased a five and a half acre tract of land and met in various locations until Saarinen's church was completed.[3]

> *The whole thing, all the planes, would grow up organically into the spire*
>
> – on North Christian Church (1960) [Saarinen 1962: 90]

The main level of the two story structure rests on a massive concrete base. The lower level nestles into a landscaped berm held back to expose a moat-like light well. The building is hexagonal in plan, elongated on the east-west axis. Six steel legs clad in lead-coated copper secure a pyramidal slate roof, then converge at an apex, rise to a tapered 192-foot central spire and terminate at a five-foot gold-leaf cross (fig. 1) [Knight 2008: 163; Thayer 1999: 4].

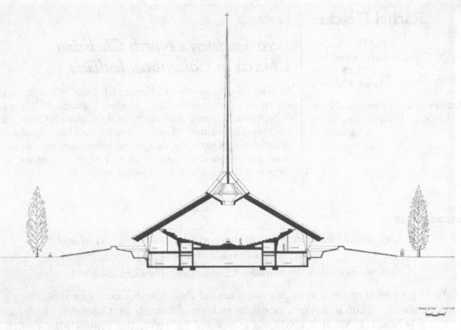

Fig. 1. North Christian Church, Columbus, Indiana. Transverse section. Image:
Eero Saarinen, Collection Manuscripts & Archives, Yale University
(Image No. 9733)

> *...put only the sanctuary above ground and make it the significant visual
> and architectural thing.*
> – on North Christian Church (1960) [Saarinen 1962: 88]

Only the main sanctuary and a small baptismal chapel, both surrounded by an ambulatory, remain above ground. Support facilities consolidate within a single unit hollowed out in the basement below. The surrounding six-foot high berm conceals from view everything but the roof and entrance. The result is a simple and distinctly singular structure that appears to hover lightly above the ground, an illusion enhanced by recessed exterior glass walls set back some twelve feet from the roof's edge [Thayer 1999: 4].

Interior

> *...you should have to work for it and it should be a special thing.*
> – on North Christian Church (1960) [Saarinen 1962: 88]

Acts of entry and passage simulate a spiritual journey to another world. From the parking area, one ascends, and then descends before reaching the building at ground level. Once inside, steeper steps lead to the sanctuary.

> *...communion is a very important act and the congregation participates in it.*
> – on North Christian Church (1960) [Saarinen 1962: 88]

Each building level follows the same hexagonal plan. Above, the main east entrance leads through a spacious vestibule to a large bowl-shaped sanctuary placed at the center and elevated to emphasize its liturgical importance. At the very center, beneath an oculus and raised on a dais, a communion table, containing twelve places arranged in two rows with a taller place at one end, represents Christ and His twelve disciples. The dais can be rotated and repositioned on other occasions.

> *...everyone feels equal and joined together.*
> – on North Christian Church (1960) [Saarinen 1962: 88]

Surrounding the communion table on five sides, mahogany benches bring congregants close to the service. Organ, choir benches and pulpit complete the round on the sixth side.

> *...immersion should be given more dignity...viewed only by family and close friends.*
> – on North Christian Church (1960) [Saarinen 1962: 90]

Free-standing walls on diagonal axes frame a secondary baptistery chapel at the west end, where immersion baptism is performed in a small hexagonal pool of white tile set into the floor. Seating faces inward toward the center. On other occasions, the pool is covered.

> *...put all that activity downstairs. Maybe underground, hidden away...*
> – on North Christian Church (1960) [Saarinen 1962: 88]

Underground perimeter classrooms, offices and other support facilities face the berm, lit by ambulatory windows at the base of the church. In the center, an auditorium is placed directly below the sanctuary.

> *The primary element to create the right spiritual atmosphere would, of course, be light.*
> – on North Christian Church (1960) [Saarinen 1962: 90]

Theatrical light sources lend mystery to the sanctuary experience. A hexagonal skylight at the base of the spire focuses natural light on the communion table, an effect augmented by recessed can lights in the ceiling nearby. Along the sanctuary's perimeter, indirect natural light from ambulatory windows reflects angled ceiling surfaces, causing the roof to appear to float [Merkel 2005: 160; Thayer 1999: 5; Miller 2006: 65].

Geometric symbolism

> *...the total concept is carried down to the smallest detail.*
> – On Interior Design (1960) [Saarinen 1962: 11]

More than decoration or ornament, abstract geometric symbols convey spiritual experience and meaning. Hexagonal figures represent the Star of David, or Magen David, recognized universally as the symbol of Judaism (cf. [Fletcher 2005: 142-145]). The spire represents for the architect "a marvelous symbol of reaching upward to God," commanding the landscape in its function as *axis mundi*. The cross at its apex represents Christianity's emergence from Judaic origins through Christ's sacrifice [Miller 2006: 65]. Early design sketches contain a variety of elementary geometric shapes and symbols— triangle, cross, square, hexagon and octagon—that contribute to the finished church plan and section (fig. 2).

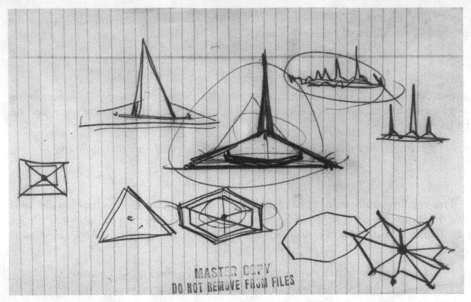

Fig. 2. North Christian Church, Columbus, Indiana. Sketches. Eero Saarinen, Collection Manuscripts & Archives, Yale University (Image No. 9736)

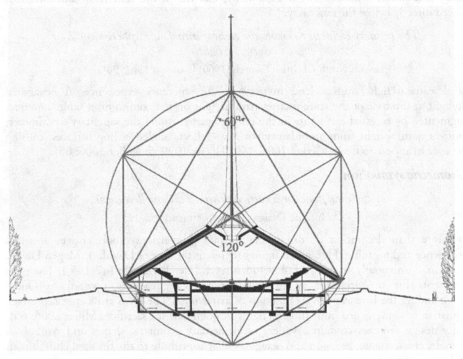

Fig. 3. North Christian Church. Transverse section with geometric overlay. Base image as in fig. 1; geometric overlay: Rachel Fletcher

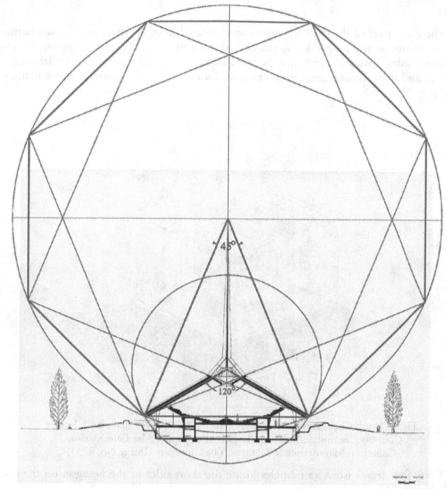

Fig. 4. North Christian Church. Transverse section with geometric overlay. Base image as in fig. 1; geometric overlay: Rachel Fletcher

Geometric analysis

Section

In section, the spire and roof converge along three equally spaced axes. In fig. 3, a regular hexagon, hexagram (or Star of David), and equilateral triangle are drawn on these axes. Roof lines coincide with two equal sides of a 120° isosceles triangle.

In fig. 4, a 45° isosceles triangle is drawn on the same base as the 120° isosceles triangle. Its apex coincides with the gold-leaf cross at the top of the spire. From the apex is drawn a circle, as shown, that encloses a regular octagon and an eight-faceted figure composed of two squares. The two squares intersect along the plane of the oculus.

Plan

The main level of the church features an elongated hexagon that encompasses berm-framed moats on the long sides, north and south, and stepped concrete approaches on the short sides, east and west. Just beyond, parallel diagonal lines delineate landscape features and trees. Inside, concentric hexagons locate the building proper and sanctuary floor (fig. 5).

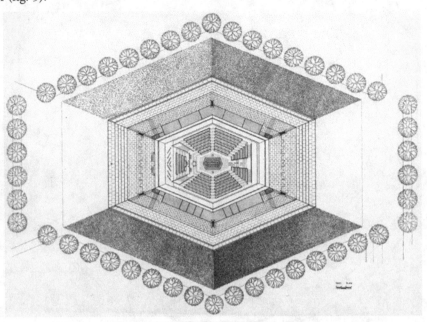

Fig. 5. North Christian Church, Columbus, Indiana. Plan of main level. Image:
Courtesy Cranbrook Archives, Maurice Allen papers. Also Eero Saarinen,
Collection Manuscripts & Archives, Yale University (Image No. 9734)

Forty-five degree isosceles triangles locate the short sides of the hexagon on the east and west (fig. 6).

Three axes, as shown, situate the extent of the hexagon and suggest the three-dimensional frame of the cube (fig. 7).

The outer hexagonal footprint derives from the root-two proportions of a regular octagon and an inscribed eight-faceted figure composed of two squares. The hexagon's short east-west sides coincide with two edges of the octagon. Its long north-south sides are equal in length to the radius of the octagon's circumscribing circle and follow the edges of the squares (fig. 8).

Saarinen's early sketches include squares and octagonal figures of a similar nature (fig. 9).

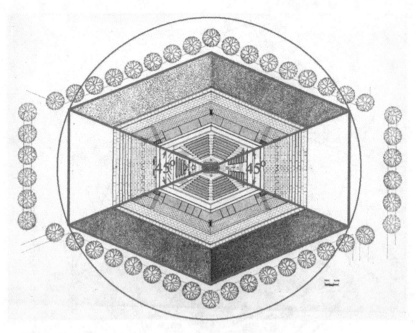

Fig. 6. North Christian Church. Plan of main level with geometric overlay. Base
image as in fig. 5; geometric overlay: Rachel Fletcher

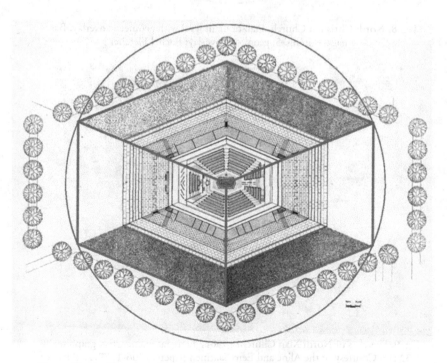

Fig. 7. North Christian Church. Plan of main level with geometric overlay. Base
image as in fig. 5; geometric overlay: Rachel Fletcher

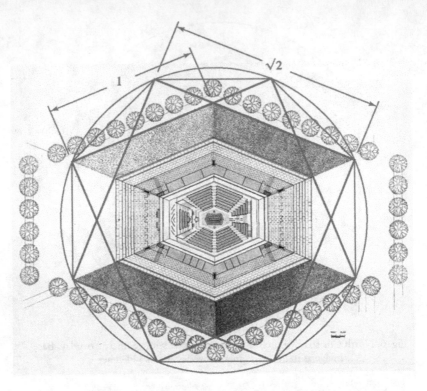

Fig. 8. North Christian Church. Plan of main level with geometric overlay. Base image as in fig. 5; geometric overlay: Rachel Fletcher

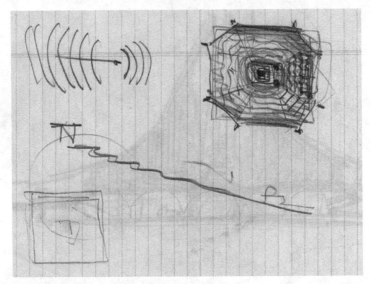

Fig. 9. Sketches of North Xian Church, 1950s / Eero Saarinen, artist, graphite; 20 x 32 cm. Courtesy of the Aline and Eero Saarinen papers, 1906-1977, Archives of American Art, Smithsonian Institution

...in any design problem, one should seek the solution in terms of the next largest thing...
– On Relationships in Design (1958) [Saarinen 1962: 11]

Within the plan, the outline of the building proper repeats and relates proportionally to the outer hexagonal footprint. Its shape follows a smaller, eight-faceted figure whose circumscribing circle inscribes the original eight-faceted figure (fig. 10).

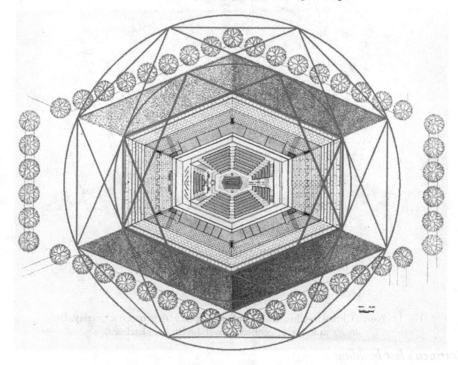

Fig. 10. North Christian Church. Plan of main level with geometric overlay. Base image as in fig. 5; geometric overlay: Rachel Fletcher

Landscape

...a building grows from its site...
– On Architecture (1959) [Saarinen 1962: 6]

Saarinen's partnership with Dan Kiley, a leading landscape architect of the modern era, ensured a seamless integration of building and setting. Kiley proposed to transform the flat treeless lot by surrounding the church with dense groupings of trees. A plot plan published in *Architectural Record* projects the building's proportions beyond the walls of the church (cf. [Saarinen's Church 1964: 187]).

The landscape plan for North Christian Church developed over several years and never materialized fully. But thickets of magnolia trees were planted along the north and south, and dwarf crabapple trees at the east and west entrances [Olivarez 2006: 274; Thayer 1999: 6, 10]. In Saarinen's early sketches, geometric lines and proportions extend beyond the church structure (cf. [Aline and Eero Saarinen Papers 1906-1977, (Image No. 5) AAA_saaralin_284659]). In the finished plan, geometric lines that delineate landscape features on the north and south parallel the long sides of the hexagonal footprint and are tangent to the circle that passes through the points of intersection of the two major squares (fig. 11).

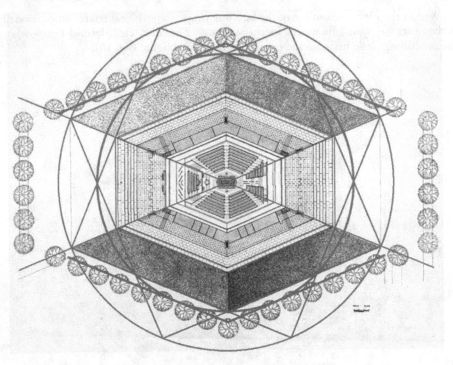

Fig. 11. North Christian Church. Plan of main level with geometric overlay. Base image as in fig. 5; geometric overlay: Rachel Fletcher

Saarinen's last building

Solving the total design of North Christian Church did not come easily to Saarinen and after two years the client grew impatient with the length of his process.[4] In April 1961, the architect responded:

> We have finally to solve this church so that it can become a great building...so that as an architect when I face St. Peter I am able to say that out of the buildings I did during my lifetime, one of the best was this little church, because it has in it a real spirit that speaks forth to all Christians as a witness to their faith [Saarinen 1962: 90].

In July 1961 Saarinen wrote that "we have finally solved the Columbus church" [Saarinen 1962: 90]. Weeks later, shortly after being diagnosed with a brain tumor, he died on the first of September.

Conclusion

...organic unity is the ideal.

– On Interior Design (1960) [Saarinen 1962: 11]

It is not certain that Saarinen composed the spaces of North Christian Church by adopting these proportions and techniques. Nor do we know if he merged hexagon and square to symbolize the Star of David united with the Cross.

Saarinen recognized the merits of geometric proportion, but in unpublished notes warned against adopting a singe canon or allowing one's sense of proportion to "to be frozen by rules and regulations" or confined to "divine and indisputable laws." Good design, he maintained, is an "integrated package" that incorporates "space, use, structure, material, texture, proportions, and last but not least, the spirit of our time" (Saarinen, "Golden Proportions," unpublished notes for a lecture, 1953 [Pelkonen 2006: 342-343]). Intuition, he proposed, is the best vehicle for capturing the total picture, more than any single element.

Intuition can be nurtured by a working knowledge of proportional techniques. These should serve a program's spiritual, social and functional requirements, and respond appropriately and organically to the situation at hand. The sublime beauty of North Christian Church demonstrates this ethic to perfection.

Notes

1. According to Jennifer Komar Olivarez, the liturgical revival begun in late nineteenth century Europe evolved through the influences of the 1938 treatise *The Church Incarnate* by German architect Rudolf Schwarz; postwar modern churches of Schwarz and German contemporary Dominikus Böhm; the 1947 German Liturgical Commission, and in the Catholic Church, Second Vatican (Vatican II) reforms adopted in 1965 [Olivarez 2006: 267-268]; see also [Ray 2005: 3].
2. Miller represented the new church, acting as chair of the search committee that selected the architect among several nationally known candidates.
3. [Knight 2008: 45; Merkel 2005: 158-159]. Architect Eliel Saarinen, Eero's father, designed First Christian Church (1939) in collaboration with his son. The National Historic Landmark Nomination locates North Christian Church near the west end of a 13.5 acre property [Thayer 1999: 4].
4. By late January 1961, Saarinen had neither resolved the lantern nor related the interior and exterior to his satisfaction [Saarinen 1962: 90].

References

Aline and Eero Saarinen Papers, 1906-1977. Archives of American Art, Smithsonian Institution. Box 2 Folder 3 Sketches of North Xian Church, Eero Saarinen 1950s. http://www.aaa.si.edu/collectionsonline/saaralin/container37465.htm

Eero Saarinen Collection. 1880-2004 (inclusive), 1938-1962 (bulk). Collection Manuscripts & Archives. Digital Images Database. New Haven: Yale University. http://images.library.yale.edu/madid/

FLETCHER, RACHEL. 2005. Six + One. *Nexus Network Journal* **7**, 1 (Spring 2005): 141–160.

KNIGHT, RICHARD. 2008. *Saarinen's Quest: A Memoir.* San Francisco: William Stout Publishers.

MERKEL, JAYNE. 2005. *Eero Saarinen.* London: Phaidon Press Limited.

MILLER, WILL. 2006. Eero and Irwin: Praiseworthy Competition with One's Ancestors. Pp. 57-67 in *Eero Saarinen: Shaping the Future*, Eeva-Liisa Pelkonen and Donald Albrecht, eds. New Haven: Yale University Press.

OLIVAREZ, Jennifer Komar. 2006. Churches and Chapels: a New Kind of Worship Space. Pp. 266-275 in *Eero Saarinen: Shaping the Future*, Eeva-Liisa Pelkonen and Donald Albrecht, eds. New Haven: Yale University Press.

PELKONEN, Eeva-Liisa and Donald Albrecht, eds. 2006. New Haven: Yale University Press.

RAY, MAIA LEA. 2005. Eero Saarinen: Creating Sacred Space. Master Thesis, University of Louisville.

SAARINEN, ALINE B. 1962. *Eero Saarinen on His Work*. New Haven: Yale University Press.

Saarinen's Church. 1964. *Architectural Record* **136** (September 1964): 185-190.

THAYER, LAURA, Louis Joyner and Malcolm Cairns. 1999. "National Historic Landmark Nomination: North Christian Church." United States Department of the Interior, National Park Service. http://www.nps.gov/history/nhl/designations/samples/in/nchrist.pdf

About the author

Rachel Fletcher is a geometer and teacher of geometry and proportion to design practitioners. With degrees from Hofstra University, SUNY Albany and Humboldt State University, she was the creator/curator of the museum exhibits "Infinite Measure," "Design by Nature" and "Harmony by Design: The Golden Mean" and author of the exhibit catalogs. She is an adjunct professor at the New York School of Interior Design. She is founding director of the Housatonic River Walk in Great Barrington, Massachusetts, co-director of the Upper Housatonic Valley African American Heritage Trail, and a director of Friends of the W. E. B. Du Bois Boyhood Homesite. She has been a contributing editor to the *Nexus Network Journal* since 2005.

R. Balasubramaniam

Department of Materials and
Metallurgical Engineering
Indian Institute of Technology
Kanpur 208 016, INDIA
bala@iitk.ac.in

Keywords : Humayun's tomb,
Taj Mahal, modular design,
Arthasastra, Indian architecture,
Mughal period, metrology

Research

On the Modular Design of Mughal Riverfront Funerary Gardens

Abstract. The modular designs of two significant funerary gardens of the Mughal period, the Humayun's tomb and Taj Mahal complexes, have been analyzed. The inherent symmetry in the designs is made evident through an analysis of the dimensions in terms of units mentioned in the *Arthasastra*, in particular the *dhanus* (D) measuring 108 *angulams* and *vitasti* (V) measuring 12 *angulams*, with each *angulam* taken as 1.763 cm. The low percentage of errors between predicted and actual dimensions has confirmed, for the first time, that the modular designs of these Mughal funerary gardens were based on *Arthasastra* units. A novel mathematical canon to analyze the dimensions of Mughal architecture has been set forth.

Introduction

There is great interest in understanding technical aspects of Islamic architecture in India, with particular emphasis on the mathematical systems used in the design and planning of Islamic structures. The geometry of the multiple axes as well as the geometry of ratios that were used in the design is of interest. Basic to all these, is the understanding of the units of measure to which Islamic structures of the subcontinent were designed and finally constructed. The Mughal period, extending for about 200 years from 1526 A.D., is a significant period of the Indian subcontinent. In this present paper attention will be focused on Mughal architecture, because presumably the elements of earlier Islamic architecture of India were well reflected in significant Mughal structures.

Most Mughal architectural designs of palace and tomb complexes follow certain patterns [Asher 1992: 19-251; Koch 2006]. The plans and designs of several Mughal structures are known [Nath 1982-2005].

The basic aim of this article is to analyze the modular architectural design of two important Mughal structures, the Humayun's tomb complex in New Delhi [Misra and Misra 2003: 15-70] and the Taj Mahal complex in Agra [Koch 2006], in order to understand the measurement units to which these structures were conceived and constructed. Ideas about metrology of Mughal structures can be only gleaned from actual structures because there are no technical manuals on architecture and building art form in the Islamic literature of the Mughal period [Koch 2006].

The two complexes chosen for this present study are good examples of the classical form of well-planned riverfront funerary gardens of the Mughal period. There is one basic inherent difference in the overall designs of these complexes, especially the relative placing of the tomb with respect to the garden. While the tomb structure is centrally located in the garden of Humayun's tomb complex, the Taj Mahal mausoleum is placed on a riverfront terrace that is located to the north of the garden. A brief introduction to the garden types of the Mughal period will provide the background about the significance of these two complexes.

Nexus Network Journal 12 (2010) 271–285   NEXUS NETWORK JOURNAL – VOL.12, No. 2, 2010 **271**
DOI 10.1007/s00004-010-0014-8; *published online* 9 February 2010

Mughal gardens

The first type of Mughal garden is the terrace garden, like the ones the Mughals developed at Kabul and Kashmir. This was a Central Asian concept, wherein the garden was laid out on a slope, blending with the landscape of the region. The main buildings were arranged on ascending terraces, placed symmetrically with respect to the central axis. This axis was usually defined by a channel sunk in a paved walkway through which collected water from a spring flowed. These individual terraces sometimes contained the canonical four-part garden, as in the case of Shalimar garden of Kashmir.

The second type of Mughal garden is the traditional *charbag* garden. The *chahar bagh* (meaning, in Persian, "four garden") or its abbreviated form *charbag* designated a cross-axial four-part garden. This kind of garden was in vogue in South Asia from even much earlier times. For example, the large royal gardens of Sigiriya in Sri Lanka were laid out in the fifth century A.D. in a cross-axial pattern [Bandaranayake 2000: 1-36].

An important point is the location of the built structure within the garden complex. The tombs were usually centrally located. This type of modular design with a centrally located tomb in the *charbag* garden finds the greatest expression at the great imperial mausoleums of Humauyn (1530-1539 A.D. and 1555-1556 A.D.) at New Delhi, Akbar (1556-1605 A.D.) at Agra and Jehagir (1605-1628 A.D.) at Lahore. While all these structures form part of funerary garden complexes, the Humayun's tomb complex will be taken up for detailed analysis because this is a riverfront complex. The plan of Humayun's tomb complex is shown in fig. 1.

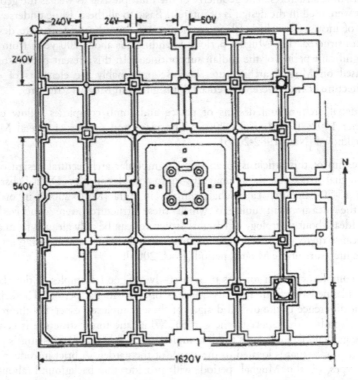

Fig. 1. Dimensions of overall complex of Humayun's tomb at New Delhi. The *vitasti* (V) equals 12 *angulams* and each *angulam* measures 1.763 cm.

There was a variation noted in some riverside Mughal constructions, which was necessitated by the luxuriant rivers available in India. Here, the building was shifted to one side of the garden such that it faced the river. This theme finds its grandest expression in the terraced riverfront *charbag* funerary garden of the Taj Mahal complex of Shah Jahan (1628-1656 A.D.). This development of riverfront gardens in Mughal India has been explained in great detail elsewhere [Koch 1997]. The plan of the Taj complex in fig. 2 shows that the main building was placed on a rectangular terrace (labelled T) running along the riverfront.

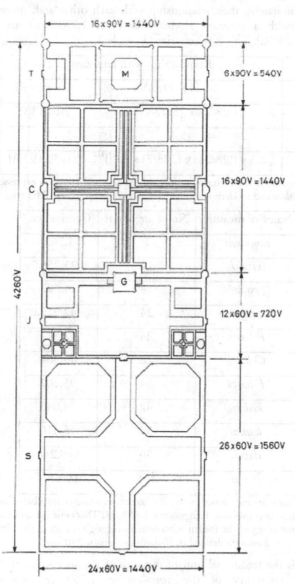

Fig. 2. Dimensions of overall complex of Taj Mahal in Agra. The *vitasti* (V) equals 12 *angulams* and each *angulam* measures 1.763 cm.

Metrology during the Mughal period

It is now important to focus on the metrology (study of the science of measurement) of the Mughal period because the analysis of modular design of Mughal funerary gardens will be illuminating only if the original units of measurements to which the structures were conceived and constructed are understood.

New insights have been obtained recently on the linear measurement units of the Mughal period [Balasubramaniam 2008a, 2009a, 2009b]. The measures used by different Mughal monarchs, their relationship with each other and, more importantly, their relationship with a constant *angulam* unit (of 1.763 cm) are now known [Balasubramaniam 2008a]. This is summarized in table 1.

Mughal Unit	Relationship with *angulam*		Measure
Sikandari gaz	28/24 x 39 x A	(=45.5 A)	80.217 cm
Illahi gaz	28/24 x 288/7 x A	(= 48A)	84.624 cm
Padshahi zira	45 A		79.335 cm
Shahjahani dira	$[2/3 \times ((28/24) \times (288/7) \times A)^2]^{1/2}$	(=39.192 A)	69.096 cm

Table 1. Linear measures of the common units used in the Mughal period expressed in terms of *angulam* and modern centimeters. Each *angulam* (A) is 1.763 cm.

Name of measure	No of *angulams*	Centimeters
angulam	1	1.763
vitasti	12	21.256
pada	14	24.682
aratni	24	42.312
P-hasta	24	42.312
C-hasta	28	49.364
F-hasta	54	95.202
kishku	42	74.046
kamsa	32	56.416
danda	96	169.248
dhanus	108	190.404

Table 2. Units of measure mentioned in the *Arthasastra* in terms of number of *angulams* and centimeters, using the conversion 1 *angulam* = 1.763 cm. This table is fundamental to the understanding of metrology of the Indian subcontinent through the ages. The different types of *hastas* are defined in [Balasubramaniam 2009a].

The *angulam* is the traditional unit of measure of the subcontinent. Recent studies have shown the constancy of the *angulam* at 1.763 cm through the ages [Balasubramaniam 2008a, 2008b, 2009a, 2009b, 2009c]. Interestingly, this value for the *angulam* was obtained without any *a priori* assumptions from the designs of Harappan civilization sites [Danino 2005, 2009]. This is also directly confirmed by the appearance of a similar unit in the Lothal ivory [Rao 1955-62: II, 689-690] and Kalibangan terracotta [Balasubramaniam and Joshi 2008] scales of the Harappan civilization. In particular, the units of measurement described in Kautilya's *Arthasastra* [Shamasastry 1939; Kangle 1986: II, 138-140], dated to around 300 B.C., were used to understand the engineering plans of most engineered structures of the Indian subcontinent through the ages [Balasubramaniam 2008a, 2008b, 2009a, 2009b, 2009c], until the adoption of British units in early twentieth century. The measurements defined by the *Arthasastra* in terms of modern centimeters are listed in table 2. The significant measurements that are emphasized in this present paper are *vitasti* (= 12 *angulam*), *dhanus* (=108 *angulam*) and *rajju* (=10 *dhanus*).

Modular design of overall complexes

Before focusing on individual structures, we will first address the modular design of the overall layout of the two complexes. The symmetric modular design is evident when the known measurements of the Humayun's tomb and Taj Mahal complexes are converted to the traditional units mentioned in Kautilya's *Arthasastra*. The plans of these two complexes in terms of traditional units of measurement are presented in figs. 1 and 2. Notably, the measures of major dimensions appear as logical numbers when expressed in terms of traditional units, in particular the *vitasti* (V). A brief discussion on the modular plans of these two complexes in terms of traditional units follows.

Humayun's tomb complex

The boundary wall of the Humayun's tomb complex is missing on the eastern side. The Yamuna flowed past this side in earlier times. The modular planning of the Humayun's tomb complex is clearly evident in the ordering of the small garden plots around the central tomb (see fig. 1). Based on the dimensions of the complex reported by Zafar Hasan [1997], the entire complex can be analyzed in terms of the units mentioned in *Arthasastra*.

It can be seen that each side of the complex measures 1620V in length. Specifically, the modular design can be understood in terms of grids of size 240V x 240V (see fig. 1). Each square garden plot measures to this grid. The tomb structure has been designed so that it is located in the central square grid of 540V x 540V (see fig. 1). The walkways are another characteristic of the cross-axial garden. The central walkway is 60V in width, while the smaller ones are 24V wide. The width of the lane bordering the wall is 12V.

In order to highlight the modular planning of the garden of Humayun's tomb, one of the quadrants of the complex has been analyzed (fig. 3). Each side of the garden plots is 240V in length. Therefore, considering one quadrant of the overall plan, the division of the length of side would be 12V+240V+24V+240V+24V+240V+30V = 810V. The last factor 30V is half of the central walkway of width 60V. One can also view the dimensions of Humayun's complex in terms of the larger *Arthasastra* unit of *dhanus*. The overall length of each side of Humayun's tomb complex in terms of *dhanus* units is 180D. The central square, where the platform is located, occupies an area 60D x 60D.

The dimensions proposed in figs. 1 and 3 and the actual measures [Hasan 1997: 117-124] are quite close, as shown by error analysis, the results of which are presented in Table 3. The error is defined as the deviation of the proposed measure from the actual measure, expressed in percentage of the proposed measure. The low errors confirm that the modular planning of Humayun's tomb complex was based on *Arthasastra* units.

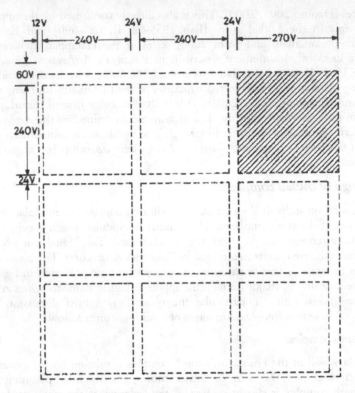

Fig. 3. Modular layout of one quadrant of Humayun's tomb complex. The *vitasti* (V) equals 12 *angulams* and each *angulam* measures 1.763 cm.

No	Location	Proposed measure		Actual measure (cm)	% error
		vitasti (V)	cm		
1.	N-S length of complex	1620V	34272.72	34853.88	-1.70
2.	E-W length of complex	1620V	34272.72	34152.84	0.35
3.	Width of main causeway	60V	1269.36	1264.92	0.35
4.	Width of minor causeway	24V	507.744	518.16	-2.05
5.	E-W and N-S lengths of central platform	540V	11424.24	11277.60	1.28
6.	E-W and N-S lengths of second platform	450V	9520.20	9326.88	2.03
7.	E-W and N-S lengths of mausoleum plinth	240V	5077.44	5029.20	0.95

Table 3. Comparison of proposed and actual dimensions [Hasan 1997] of some important sections of Humayun's tomb complex. The vitasti (V) equals 12 angulams and each angulam measures 1.763 cm.

Taj Mahal Complex

The overall plan of the Taj Mahal complex is shown in fig. 2. The complex can be considered to consist of four sections. The riverfront terrace (T) is the northernmost section. The garden (C) is located beside the terrace. The forecourt (J) and caravanserai (S) on the south complete the complex.

The dimensions of significant locations and features of the Taj complex are given in [Barraud 2006a; 2006b]. Therefore, the dimensions of the complex can be determined in *Arthasastra* units. The total length of the complex, including the forecourt and the caravanserai is 4260 V. Each of the four sections of the Taj Mahal complex have been designed based on grids. The modular planning of the four sections of the Taj Mahal complex has been described in great detail elsewhere [Balasubramaniam 2009b]. The terrace and *charbag* sections have been designed and planned to a larger grid size than that used for the forecourt and caravenserai sections.

The modular planning of the terrace and garden section of the Taj complex are highlighted in the schematic of fig. 4. The unit mentioned in this figure is the *dhanus* (D), each being equal to 108 *angulam*. This can be converted to *vitasti* (V), which is equal to 12 *angulam*. Therefore, 1D = 9V.

The total length of the riverfront terrace and the garden is 220D. The garden occupies an area of square 160D x 160D. The length of the riverfront terrace is 60D. Further, the garden has been planned by dividing the total garden area into smaller grids of size 10D x 10D, as shown in fig. 4.

It is interesting to note that 10D is equal to 1 *rajju*, a measure mentioned in the *Arthasastra* for measuring larger distances [Shamasastry 1939; Kangle 1986: II, 138-140]. (Incidentally, *rajju* in Sanskrit means "rope.") Therefore the modular planning of the garden and riverfront terrace sections of the Taj complex can be understood using a modular grid pattern of 1 *rajju* x 1 *rajju* (= 10D x 10D = 90V x 90V).

The errors between the proposed and actual dimensions [Barraud 2006a; 2006b] for the Taj complex are low (see table 4). This confirms the use of *Arthasastra* units for the modular planning of the Taj complex.

No	Location	Proposed measure		Actual measure (cm)	Percentage error
		vitasti (V)	cm		
1.	E-W length of terrace	1440V	30464.64	30084	+1.25
2.	N-S length of terrace	540V	11424.24	11189	+2.06
3.	E-W and N-S lengths of marble platform	450V	9520.20	9569	-0.51
4.	E-W and N-S lengths of mausoleum plinth	270V	5712.12	5690	+0.39
5.	E-W and N-S length of *charbag*	1440V	30464.64	29631	+2.74
6.	E-W length of *jilaukhana*	1440V	30464.64	30084	-1.25
7.	E-W length of caravanserai	1440V	30464.64	30084	-1.25
8.	N-S length of caravanserai	1560V	33003.36	33490	-1.47

Table 4. Comparison of proposed and actual dimensions [Barraud 2006a; 2006b] of some important sections of Taj Mahal complex. The *vitasti* (V) equals 12 *angulams* and each *angulam* measures 1.763 cm.

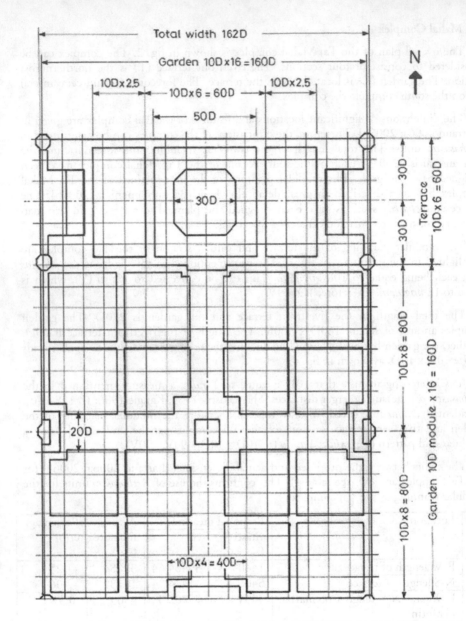

Fig. 4. Modular layout of riverfront terrace and charbag of Taj Mahal. The *vitasti* (V) equals 12 *angulams* and each *angulam* measures 1.763 cm.

The above analysis has highlighted the important fact that the overall design of the Humayun's tomb and Taj Mahal complexes can be rationalized in terms of the traditional length units of the subcontinent, especially the ones defined in *Arthasastra* with the *angulam* taken as 1.763 cm. This is an important original finding of this study. This concept will be now applied in order to understand the modular design of the platforms on which the two mausoleums are erected.

Modular planning of the platforms

The positioning of the two mausoleums with respect to the platforms on which they stand is of interest.

In the case of the Humayun's tomb, the central feature of the garden is a platform that is 540V square with corners slightly cut off. It is paved with large blocks of Delhi quartzite and is about 4 feet tall from ground level. In the centre of this platform is another platform, 450V square and chamfered at the corners. This stands about 22 feet (approximately 30V) above the level of the lower platform. The relative positioning of these two platforms and the mausoleum of Humayun is shown in fig. 5. Each face of the second platform contains 17 arched recesses, 8 on each side of the central steps and one recess in the splayed corners of the platform. The second platform serves as the base for the mausoleum. Steps in the centre of each of the four sides lead to the actual tomb on the second platform. The mausoleum stands on a low plinth in the center of the upper platform and is entered from the south. Each side of the square of mausoleum plinth is 240V. All these dimensions have been marked in fig. 5. The match between proposed and actual measurements [Hasan 1997: 117-124] for the length of the two platforms and the mausoleum plinth is excellent (see table 3).

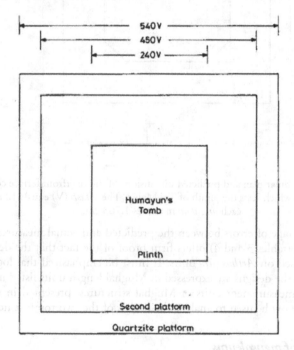

Fig. 5. Proposed modular plan and predicted dimensions of the central platform on which stands the Humayun's tomb. The *vitasti* (V) equals 12 *angulams* and each *angulam* measures 1.763 cm.

In a similar manner, the positioning of the Taj mausoleum in the riverfront terrace is as shown in fig. 6. The relative dimensions of the platform and the plinth of the Taj, and their relation to the N-S length of the riverfront terrace (540V) are noteworthy. The length and breadth of the marble platform is 450V, while the length and breadth of the plinth of the mausoleum equals 270V (see fig. 6). The match between predicted and actual measurements [Barraud 2006a; 2006b] for the marble platform (450V) and the mausoleum plinth (270V) is excellent (see table 4). In this manner, the square (representing the plinth of the mausoleum) of 270V per side was planned in the centre of another square (representing the marble platform) of 450V per side. Incidentally, these dimensions are symmetrically related to the N-S length of the terrace, which equals 540V. The overall symmetry of this design scheme can be appreciated in the plan shown in fig. 6.

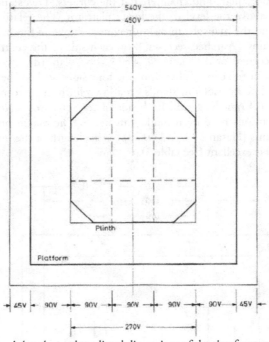

Fig. 6. Proposed modular plan and predicted dimensions of the riverfront terrace containing the marble platform on which rests the plinth of Taj Mahal. The *vitasti* (V) equals 12 *angulams* and each *angulam* measures 1.763 cm.

The low percentage of errors between the predicted and actual measures of platforms of these complexes (tables 3 and 4) offers firm proof of the fact that the designs of these complexes were based on *Arthasastra* units. It must be emphasized that logical numbers do not result when the designs are expressed in Mughal length units listed in table 1. The novel proposal of measurement units of Mughal structures, presented in this paper for the first time, will now be used to analyze the design of the two tomb structures in these two complexes

Modular design of mausoleums

The favorite Mughal form for mausoleums is a centrally planned building known in Persian as *hasht bihisht* or eight paradises [Koch 2002: 45-50]. The *hasht bihisht* design appears dramatically in the Humayun's tomb (see fig. 7) and Taj Mahal mausoleum (see fig. 8). In both the tombs, four radially planned *hasth bihist* elements can be noted. Both the cross-axial pattern as well as a diagonal axis pattern is evident. The modular designs can be better appreciated by discussing the metrology of these structures individually.

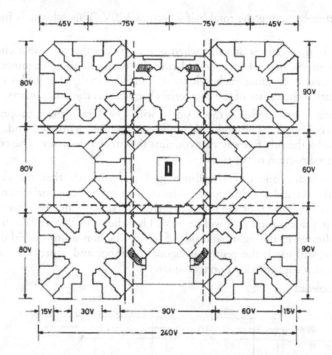

Fig. 7. Proposed modular plan and predicted dimensions of Humayun's mausoleum. The *vitasti* (V) equals 12 *angulams* and each *angulam* measures 1.763 cm. Notice the different manner in which the length of each side of the structures has been divided.

Humayun's Mausoleum

Analysis of the dimensions of the tomb revealed its association with the traditional units of *Arthasastra*. Each side of Humayun's tomb is 240V. One can then divide each side in thirds, to give nine grids with each grid of size 80V x 80V. The grid based on this division is marked with dark lines in fig. 7. However, in this case, this may not be an appropriate grid size to appreciate its design because the finer dimensions of the structure cannot be reconciled by intricate subdivisions of the 80V unit. The dimensions of significant features like the width of doorways, etc. are not based on smaller subunits of 80V.

Careful examination of relative dimensions of various features of Humayun's tomb reveals that a grid pattern of 15V to the side better reflects the symmetry of the structure. Most of the finer modular sub-divisions can be understood by considering 15V divisions. For example, considering one side of the tomb, the length of the corner chamfer is 15V. The distance from the chamfer to the small doorway is 15V, with the doorway occupying a width of 30V. The width of the wall adjoining this small side doorway is 15V. Immediately after is the central arched doorway whose entire width is 90V. The main doorway entrance, located inside the doorway, is 60V. Similarly, the width of the entire side doorway section containing the side doorway is 60V.

Several other division schemes can be used to highlight the relative positioning of major structural features. Each side can first be simply divided into two halves (of 120V) with each half further being divided as 75V from the centre line of the central doorway to the centre line of the side door. Further, the distance from this location to the extreme corner is 45V, thereby giving the total half length as 120V. This scheme is indicated in the top of fig. 7.

Another division scheme is at work when one considers the octagonal chambers in the four corners of the mausoleum. Considering the centre of these chambers, another grid can be generated dividing each side as 45V+150V+45V. This grid will have its intersection points at the centre of the octagonal chambers on the four corners.

Another way to divide the side will be in the proportion of 15V+60V+90V+60V+15V. The end 15V units are for the chamfered end. The 60V length is covered by the side faceway that contains the smaller doorway. The central 90V occupies the central portion of the tomb.

In another scheme, one can visualize a grid pattern such that the central grid coincides with the vertical and horizontal sides of the central octagonal chamber of the tomb. In that case, the width of the central square will be 60V and therefore the side of the grids adjoining this central grid will be 90V. The scheme of 90V+60V+90V division of the side is indicated in the right hand side of fig. 7. The non-uniform grid drawn with these dimensions will mark the central octagonal chamber and moreover, each corner chamber will be now aligned with a perfect square.

Taj Mahal mausoleum

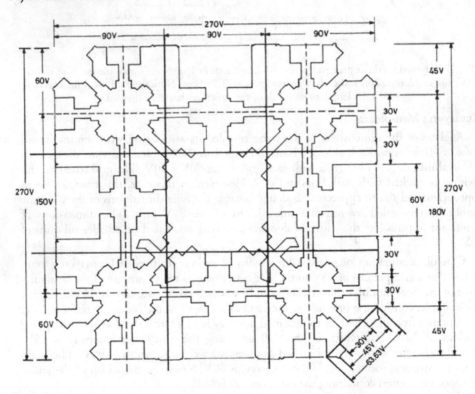

Fig. 8. Proposed modular plan and predicted dimensions of Taj Mahal mausoleum. The *vitasti* (V) equals 12 *angulams* and each *angulam* measures 1.763 cm. Notice the different manner in which the length of each side of the structures has been divided.

The proposed modular plan and dimensions of different sections of the mausoleum are seen in fig. 8. In this figure, the *vitasti* (V) has been used. First, it is confirmed that the Taj Mahal was designed using a master square of 270V to the side (see fig. 8). The match of the predicted values of the dimension of the each side of the plinth with actual measured values [Barraud 2006a; 2006b] is excellent (table 4). The usefulness of the logical number 270 (= 3^3x10) for accomplishing further subdivisions is highlighted in [Balasubramaniam 2009b]. The multiplication factor 10 involved in this number 270 facilitated decimal division of dimensions, which was important considering the intricate inlay and exquisite mosaic work on the walls and floor of the Taj Mahal. It is important to emphasize that the possibilities for the important triadic division of space are evident *only* when the dimensions of the Taj Mahal are considered in units of *vitasti*.

It is interesting to explore how the plan of the Taj was divided. In the first instance, the plan can be divided into nine smaller squares of side 90V (see fig. 8). The tripartite composition and triadic symmetry of the Taj structure can be readily appreciated by considering this grid division. The logical number of 270V for the length of the side was divided into squares of 90V x 90V of the *hasth bisith* design. Further subdivision of the 90V length into thirds is evident in the width of the large arched doors (60V) and the small arched doors (30V) on each (outer) face of the mausoleum (see fig. 8). The predicted value for the width of the large door is 1269.36 cm. (=60V) and this is only −1.70% away from the average measured value of 1291 cm. [Barraud 2006a; 2006b].

The concept of triadic division can be extended further by considering division of 90V into three 30V segments. Further, these 30V units can be subdivided into three divisions of 10V each. It is reasonable to propose that this triadic division of space was very critical to and aided in the tripartite composition of the Taj Mahal.

The lengths of several major sections and architectural elements of the Taj Mahal reveal that the side of length 270V was also divided into other different schemes, like 60V+150V+60V or 45V+180V+45V. These possibilities have been marked in fig. 8. The division of the 270V side into unequal lengths of 60V+150V+60V results in the grid pattern whose intersection points precisely match the centers of the octagonal chambers on the four corners. This grid pattern has been shown by a dotted line in fig. 8. Using the scheme of dividing the 270V length into lengths of 45V+180V+45V, it is possible to recognize the design of the chamfered corners of the Taj Mahal. Considering each corner square of dimensions 45V x 45V, it is first noted that the length of the corner section does not correspond to the length of the diagonal of this square (namely $(45/\sqrt{2})V$ = 63.63V). The corner has been designed such that its length is 30V and the length of the corner door archway is 20V (see fig. 8).

The important finding of the present study is that the modular design and inherent symmetry of Humayun's tomb and the Taj Mahal be appreciated only when the dimensions are expressed in *Arthasastra* units. This implies the utilization of traditional building principles of the subcontinent in the construction of these two Mughal masterpieces. The continuity of traditional engineering technology of the subcontinent is therefore highlighted by the analysis presented in this study. It is confirmed that there is no drastic change in the units of measurement used in architectural constructions during the Mughal period. The native artisans and designers presumably followed the standard that was well known to them over a long period of time [Balasubramaniam 2008a, 2008b, 2009a, 2009b, 2009c].

Conclusions

The modular design and inherent symmetry of two significant Mughal funerary garden constructions, Humayun's tomb and the Taj Mahal, have been revealed by analyzing their dimensions based on traditional Indian units of measure mentioned in the *Arthasastra*, in particular the *dhanus* (D) measuring 108 *angulams* and *vitasti* (V) measuring 12 *angulams*, with each *angulam* taken as 1.763 cm. The low percentage of errors between the predicted and actual dimensions of different sections of these monuments confirmed the novel approach to metrological analysis of Mughal structures.

Acknowledgment

The author thanks the Archaeological Survey of India for support of his studies on Humayun's Tomb in New Delhi and Taj Mahal complex in Agra.

References

ASHER, C.B. 1992. *Architecture of Mughal India*. New Cambridge History of India. Cambridge University Press.

BALASUBRAMANIAM, R. 2008a. New Insights on Metrology during the Mughal Period. *Indian Journal of History of Science* 43: 569-588.

———. 2008b. On the Mathematical Significance of the Dimensions of the Delhi Iron Pillar. *Current Science* 95: 766-770.

———. 2009a. On the Confirmation of the Traditional Unit of Length Measure in the Estimates of Circumference of the Earth. *Current Science* 96: 547-552.

———. 2009b. New Insights on the Modular Planning of the Taj. *Current Science* 97: 42-29.

———. 2009c. New Insights on Metrology during the Mauryan Period. *Current Science* 97: 680-682.

BALASUBRAMANIAM, R. and J.P. JOSHI. 2008b. Analysis of Terracotta Scale of Harappan Civilization from Kalibangan. *Current Science* 95: 588-589.

BANDARANAYAKE, S. 2000. *Sigiriya: City Palace and Royal Gardens*. Colombo: Central Cultural Fund, Ministry of Cultural Affairs.

BARRAUD, R. A. 2006a. Modular Planning of the Taj. Pp. 108-113 in E. Koch, *The Complete Taj Mahal: And the Riverfront Gardens of Agra*. London: Thames & Hudson.

BARRAUD, R. A. 2006b. Factfile. Pp. 258-259 in E. Koch, *The Complete Taj Mahal: And the Riverfront Gardens of Agra*. London: Thames & Hudson.

DANINO, M. 2005. Dholavira's Geometry: A Preliminary Study. *Puratattv,* 35: 76-84.

DANINO, M. 2009. New insights into Harappan town-planning, proportions, and units, with special reference to Dholavia. *Man and Environment* 33: 66-79.

HASAN, Z. 1997. *Delhi Zail*. Zail. Pp. 117-124 in Vol. III of *Monuments of Delhi: Lasting Splendour of the Great Mughals and Others*. New Delhi: Aryan Books International.

KANGLE, R. P. 1986. *The Kautilya Arthasastra*. New Delhi: Motilal Banarsidass.

KOCH, E. 1997. The Mughal Waterfront Garden. Pp. 140-160 in *Gardens in Times of Great Muslim Empires: Theory and Design, Muquarnas*, Supplement, Vol. 7, A. Petruccioli, ed. Leiden: E. J. Brill.

———. 2002. *Mughal Architecture: An Outline of Its History and Development (1526-1858)*. New Delhi: Oxford University Press, New Delhi.

———. 2006. Chapter II: The Construction of the Taj Mahal. Pp. 83-101 in *The Complete Taj Mahal: And the Riverfront Gardens of Agra*. London: Thames & Hudson.

MISRA, N. and T. MISRA. 2003. *The Garden Tomb of Humayun: An Abode of Paradise*. New Delhi: Aryan Books International.

NATH, R. 1982-2005. *History of Mughal Architecture*, 4 vols. New Delhi: Abhinav Publications.

RAO, S. R. 1955-1962. *Lothal A Harappan Port Town*. New Delhi: Manager of Publications, Government of India Press.

SHAMASASTRY, R. 1939. *Kautilya's Arthasastra* (trans.), 3rd ed. Mysore: Mysore Printing and Publishing House.

About the author

R. Balasubramaniam is a full professor in the Department of Materials and Metallurgical Engineering at the Indian Institute of Technology, Kanpur where he teaches courses on materials science and engineering, corrosion, surface coatings technology and history of metallurgy. He earned a Ph.D. in Materials Engineering at the Rensselaer Polytechnic Institute of Troy, USA in 1990 and his research activity is based on corrosion, materials-hydrogen interactions and Indian archaeometallurgy. In the recent past, he has actively researched the metrological tradition of the Indian subcontinent with emphasis on the unit of measurement used through the ages. The widely-published scholar is the author of nine books and serves on the editorial board of several international scientific journals.

Maryam Ashkan

Department of Architecture,
Faculty of Built Environment,
University of Malaya
Lembah Pantai
50603 Kuala Lumpur, MALAYSIA
maryamashkan@gmail.com

Yahaya Ahmad

Department of Architecture,
Faculty of Built Environment,
University of Malaya
Lembah Pantai
50603 Kuala Lumpur, MALAYSIA
yahaya@um.edu.my

Keywords: Geometric designs,
Islamic mathematics, Eastern
dome history, Middle East and
Central Asian domes,
discontinuous double-shell
domes, dome typology

Research

Discontinuous Double-shell Domes through Islamic eras in the Middle East and Central Asia: History, Morphology, Typologies, Geometry, and Construction

Abstract. This paper presents a developed geometric approach for deriving the typologies and geometries of discontinuous double-shell domes in Islamic architecture. Common geometric attributes are created using a corpus of twenty one domes that were built in the Middle East and Central Asia, beginning from the early through to the late Islamic periods. An outline of the origin and development of the discontinuous double-shell domes and their morphological features are addressed. Using the al-Kashi geometrical essences, a four-centered profile as an initial shape is constructed based on new geometric parameters to deduce the geometric commonalities of the two aspects of formal language (typologies and geometries) of such domes. Common geometric prototypes for typical profiles shared by the study cases are generated and formulated according to a proposed system. The theoretical frame work for the formal language of discontinuous double-shell dome architecture is structured to indicate a moderate development of this sort of Islamic domes and highlight the specific geometric relationship between the Islamic domical configurations and practical mathematic rules for many decades. It can also be established a basic approach for considering the geometric compositional designs and the typological derivations of the other eastern domes.

1 Introduction

It has generally been recognized that domes whether as single domical buildings or in large complexes of buildings, have played significant role in Islamic architecture. They are different considerably in sizes and types. The double-shell domes included the majority of Islamic domes and had gradually developed from the early Islamic epochs through to the late Islamic era. An Islamic or eastern double-shell dome, which is defined as the dome whose two shells have noticeable distances, is the matter of this paper.

The developed four-centered profile with newly geometric constitution, including the variances in angles, divisions, which have been relied on the al-Kashi geometrical essences, presents a novel method for deriving the diverse typologies and geometries of discontinuous double-shell domes. Using this method, geometric properties and compositions of the eastern domes can widely be considered.

The discontinuous double-shell domes increasingly demonstrate the high level of development of the Islamic dome architecture in the Middle East and central Asia. They display a formal language encompassing geometric concepts, typologies variations, and morphological constitutions. These possess the specific geometric entities of the domical formations that represent the physical link between practical mathematic and their architectural expressions in order to emphasis the meaning of 'centrality' in the Islamic architecture. The typological structure underlines the diversities of the discontinuous double-shell domes that can be geometrically analyzed and syntactically systematized for formulating a comprehensive compositional language. It also helps understand the styles and aesthetic principles of these domes in Islamic architecture.

Despite several existing studies about the Islamic domes and their relative meanings, the discontinuous double-shell domes still do not have completely known architectural morphology, typology, geometrical context, and even associated terminologies.

The study of characteristics of the discontinuous double-shell domes and geometric thoughts can be manipulated to give contemporary meanings to the traditional designs and principles of the Islamic dome styles. Also, the developed geometric method has the potential to be used analogically in analyzing and understanding the essences of different sorts of the eastern domes.

This present paper contains the following parts: 1) a brief outline of the origin and development of the discontinuous double-shell dome in historical architecture; 2) a brief elaboration of the Islamic mathematician's contributions and the al-Kashi geometry essences; 3) an exposition of derived morphologies of the discontinuous double-shell domes; 4) derivations of their typologies, geometric designs, and associated drawings of such domes according to the developed geometrical method; and 5) a brief discussion on their common construction methods and techniques.

2 Background

2.1 Historical outline of the origin and development of the discontinuous double-shell domes in historical architecture

Generally in Islamic architecture, the regional terminologies used to name the distinct building functions include *qubbat* (in Arabic), *gunbad* (in Iran and Afghanistan), *gumbaz* (in Uzbekistan), and *kumbett* (in Turkey). They are normally referred to the most distinguished forms of these domes in the Middle East and Central Asia.

After the introduction of the wooden domes in the near East, such as Sion church (456-460 A.D.) in Jerusalem and the dome of Rock (621 A.D.) before and after the coming of Islam [Smith 1971], thousands of masonry domes were built in this realm. The absence of sufficient literature and meanings to assess the existing evidences had regularly frustrated investigations about the Islamic domical styles and their architectural configurations divorcing from their ancient and pre-Islamic origins. Nevertheless, the Islamic domes certainly influenced on the Western domical architecture in the nineteenth century [Grabar 1963; 2006].

Morphologically, in the construction of the eastern domes, the shell(s) can be put together in three different ways (fig.1). These include one shell (the earliest type of the eastern domes: (OS-Type 1 and OS-Type 2), two shells and three shells [Hejazi 1997]. However, few samples of these triple shells that emerged in comparison to large numbers of the other sorts can thus verify its origin from the double-shell domes [Grangler 2004].

Regarding the double-shell types, two subdivision groups have been defined based on how these two shells are composed together. They are the continuous and the discontinuous groups.

In the continuous double-shell domes, sometimes, there exists no considerable distance between the shells (CD-Type 1), or they are connected by brick connectors (CD-Type 2), but very often the distance between these shells are small (CD-Type 3) [Hejazi 1997]. It could thus be said that the continuous two shells domes are called 'evolving' from the one shell domes to the two shells domes in the Islamic dome architecture development. The constructions of the one shell dome were continued up to the late Islamic era [Stierlin 2002].

In the discontinuous double-shell domes, there are considerable distances between the two shells. The discontinuity may start either from the base (DD-Type 3) or from the top of the drum (DD-Type 1 and 2) [Hejazi 1997]. This is considered higher than the other types of the Islamic domical typologies (DD-Type 2; TS-Type 1).

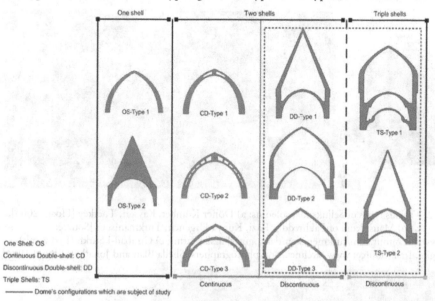

Fig. 1. Illustration of the Islamic dome typologies according to the composition of their shell(s), after [Hejazi 1997]

One of the main advantages of discontinuous double-shell domes' structure is the separation the weathering surface from the internal shell and thereby, substantially giving improved weather protection [Mainstone 2001]. Structurally, the weight of a given overall breath of construction is reduced when using the light shells. In fact, its construction method was also extremely successful, despite the seismic conditions in the Middle East and central Asia [Hejazi 2003; Farshad 1977].Architecturally, it permitted an increase in the external size and height of the dome thereby making it more imposing without the necessary increase in its internal height which improved its aesthetical meanings and splendors [Hillenbrand 1994; Michell 1978]. Mechanically, the internal shell boost sound-reflected compared to the other types of the eastern domes. Any whisper on one side of the domical chamber is easily heard because of its specific shape. This principle is also applied to all forms of energy under the internal shell [Irfan 2002].

Overall speaking, the discontinuous double-shell domes as a vast group of Islamic domes, in retrospect, resulted in the fairly continuous developments of generous proportions in dome practice in order to reach the meaning of 'centrality' in the Islamic architecture [Michell 1978].

Historically, the most constant practices were not appeared up to the eleventh century by the Seljuks architecture. Their architectural contributions and movements involved innovative characteristics of various types of mausoleums [Creswell 1958]. They showed enormous variety of appearances including octagonal, cylindrical (also called tower, see fig.2a) and square shapes (fig. 2b, 2c) topped with two types of discontinuous two shells: 'conical' roof formations (especially in Anatolia and Persia) or pointed' shapes (in Persia)[Saud 2003]. Nevertheless, the main origin of these compositions is the earlier tower tombs of that area, especially, the prominent Gunbad-e Qabus at Gurgan (Iran, 1007 A.D.) [Saud 2003].

(a)	(b)	(c)

Fig. 2. Various types of Seljuks mausoleums: a) Döner Kümbet, Kayseri, Tuerkey [Hoag 2004]; b) Mausoleum of Fakhreddin Razi, Kunya Urgench, Turkmanistan [Source: www.advantour.com/turkmenistan/dashoguz/il-arslan.htm]; c) Gunbad-i Surkh (Red tomb), Maragha, Iran [Source: www. archnet. org, photographers: Sheila Blair and Jonathan Bloom 1984]

Fig. 3. The discontinuous double-shell domes placed over tomb towers, Kharraqan, Iran [Photo: Seherr-Thoss 1968; Crosse-section after the Iranian Ministry of the Cultural Heritage (ICHO)]

In the conical roof (shell) cases, stone slabs or brick layers rested on the lower roof (shell) with some internal voids for reducing the weight and protecting the lower parts [Mainstone 2001]. In some cases, the tomb towers are topped with pointed shells. The earliest known of such discontinuous double-shell domes were a couple of eleventh century Iranian tomb towers (fig. 3). Both the internal and external masonry shells have similar thickness and profiles as well as were composed without internal connections and interconnecting wooden struts [Mainstone 2001]. Nevertheless, these tomb towers demonstrated primary efforts in designing a conflict between the external appearance of the domes and its aesthetic interior space in this period.

Another typical building typology, which showed considerable development under the Seljuk patronage and were exceedingly used by Mongols later, is the domed mausoleums were regionally built based on the developed architectonic configurations in Greater Khorasan (now divided between Iran, Afghanistan, Tajikistan, Uzbekistan, and Turkmenistan countries; see fig 4). Their structures consist of two cube-shaped stories topped with huge dissimilar two shells domes [Pope 1971]. Study of the damaged external shells of these pointed samples demonstrated the lacks of structural knowledge and information of the proportional design.

(a)   (b)

Fig. 4. Two stories Seljuks mausoleums; a) Anonymous mausoleum, Khorasan, Iran [Memarian 1988]; b) Mausoleum of Sultan Sanjar (before conservation), Merv, Turkmenistan [Hoag 2004]

After the severe architecture degeneration, which was caused by the Mongol's invasions and their successor's, Timur (that produced a gap in the dome construction evolution), the material culture of the Middle East and central Asia flourished again through Ilkhanids (in Iran) and Timurids (in Uzbekistan) [Stierlin 2002]. Architects and artists from Asia Minor, Azerbaijan, the Caucasus, India, Iran, and elsewhere were compelled to help the constructions of the often colossal state buildings for both sacred and secular purposes [Michell 1978]. Subsequently, these were characterized by the resemblance a huge of monumental discontinuous double-shell domes throughout this realm. While the Ilkhanids domes were extensively built for the funerary usages, but the Timurids domes were regularly attached to *madrasa* (religious school) and were often in pairs instead of erecting on freestanding mausoleums and mosques [Hillenbrand 1994].

In this regard, the dome over the Sultan Bakht Aqa mausoleum in Isfahan (1351-52 A.D.) was the earliest known example of the discontinuous double-shell dome where the external and internal shells substantially exhibited in different profiles with radial stiffeners, as shown in fig. 5 [O'Kane 1998]. The dome of Sultaniya complex in Cairo which were likely erected by Sultan Hasan about thirteen century, believed that is the origin of formation of the Sultan Bakht Aqa tomb. Nevertheless, the use of internal stiffeners between its two shells can be demonstrated its Persian origin [O'Kane 1998].

Consequently, the dome architectures were rapidly incorporated and altered into local styles after the Timurids epoch by appearing three specific local dynasties [Hillenbrand 1999] including: the Safavids in Iran (1501-1732 A.D.), the Shaybanids in Central Asia (1503- ca. 1800 A.D.), and the Mongols in India (1525-1858 A.D., out of scope of this paper).

Fig. 5. Sultan Bakht Aqa mausoleum with its internal stiffeners, Isfahan, Iran [Photo: Authors; Cross-section after Memarian 1988]

(a)   (b)

Fig. 6. Uniform style of Timurid domes a) Bibi Khanum mosque, 1398/1405 A.D., Samarkand, Uzbekistan, [Source: www.archnet. org; photographer: Hatice Yazar, 1990]; b) Khawaja Abu Nasr Parsa mausoleum, 1460/1598 A.D., Balkh, Afghanistan [Source: www.archnet. org; photographer: Stephen Shucart 2005]

They were dominated by the skilful use of diverse building materials (vernacular architecture) and well-developed construction techniques which existed in each region (fig.7) [Michell 1978]. At last, the application of innovative approaches in the Islamic domical constructions became less important in the Middle East and Central Asia in the middle of the late Islamic era. The constructions of Mongolian domes, however, survived longer up to the end of late Islamic era in India [Stierlin 2002].

The most common prototype of discontinuous double-shell domes consists of the external shell (the most importance component and the most visible part of dome), high drum, internal shell, and radial stiffeners within the wooden struts. The latter was used to fill the space between shells as well as integrating the whole components (fig. 8).

Fig. 7. The last generations of Islamic domes a) Tilla Kari madrasa, 1746/1660 A.D.(Shaybanids), Samarkand, Uzbekistan [Source: www.archnet.org, photographer: Arul Rewal 1990]; b) Khawaja Rabi mausoleum, 1617-22 A.D. (Safavids), Mashhad, Iran [Source: Authors]

Fig. 8. The distinct configurations of the discontinuous double-shell domes; a) Turabek Khanum tomb, Seljuks, Uzbekistan. After [Hillenbrand 1994]; b) Gur-e Amir Timur, Timurids, Uzbekistan. After [Hillenbrand 1994]; c) Shah mosque (*Masjid-e Imam*), Safavids, Iran. After [Stierlin 2002]

2.2 Historical background of the mathematicians' role in the design of Islamic domes

In an aforementioned development treatise of the Islamic discontinuous double-shell domes, the role of the mathematicians can not be overlooked. Overall speaking, the Islamic mathematic in contrast with the Greek mathematics can also be called "the mathematics of practitioners due to close relationship of theory and practice [Özdural 2000]. Its proper demonstration can be derived from the works of al-Buzjani's student who recorded contexts of his meetings with master builders and architects to discuss solutions to construction problems [Özdural 1995].

The locations of Islamic mathematic scientific centres were in the present-day Iran and Iraq. The main used languages by mathematicians for writing mathematical treatises were "Persian" and "Arabic" (principle language like Latin in Medieval Europe), and "Turkish" (more translation versions). Because of this, it is often called "Arabic mathematics". It is interesting to observe how the mathematicians had to take into consideration the master builders' objectives for using geometry in design as well as how the artisans had to realize the differences between precise and approximate approaches [Katz 2007]. In fact, the main efforts aimed to define and formulate 'an exact geometric method' rather than to determine certain proportions, not only for the dome designs, but also for the arches and vaults compositions.

The primary textual signs of their assistance can be seen in the Ismail Samanid mausoleum in Bukhara which was built in the early Islamic era (fig. 9) and its formation was entirely designed by the geometrical principles of three celebrated mathematics: al-Khorezmi, al-Fargani, and Ibn-Sino [Askarov 2007; Pope 1971].

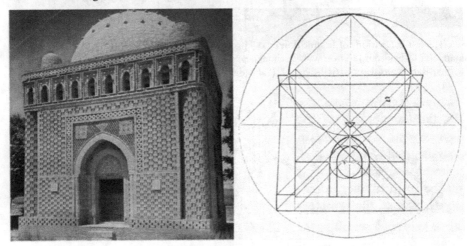

Fig. 9. Geometric design of configuration of Ismail Samanid mausoleum, Bukhara, Uzbekistan [Pope 1976]

A descriptive geometry method was used by Abu Sahl al-Quhi (circa. 1000) for projecting circles on the sphere into the equatorial plane. Then, he rendered them back onto the sphere in an outstanding visual manner. However, it would seem that he was not interested in the practical mathematics [O'Connor and Roberstson 1999]. Another known text is *Kitab fi ma yahtaj ilayh al-kuttab wa'l-ummal min 'ilm al-hisab* (a book on the geometric constructions necessary for craftsmen), which was written by Abu'l- Wafa Buzjani (circa. 1000) for practical use. Although neither the specific domical geometry nor the related information was mentioned in this book, his particular geometric methods have helped solve problems by simply using a ruler and a fixed compass [Jazbi 1997]. The long absence of developments during ca. 1000 until ca. 1400 was the result of the genocide of scientists by Mongols and Timur troops [Stierlin 2002].

The Buzjani method was developed by the celebrated al-Kashi mathematician (1390-1450 A.D.) to create the practical geometry for a variety of dome constitutions. Ghiyath al-Din Jamshid Mas'ud Kashani (al-Kashi) ranks among the greatest mathematicians and astronomers in the Islamic world. By far his extensive book is *Key of Arithmetic* (Book IV), on "Measurements" where the last chapter, 'measuring structures and building', was written for practical purposes by using geometry as tools for his calculations [Dold-Samplonius, 1992, see fig. 10b].

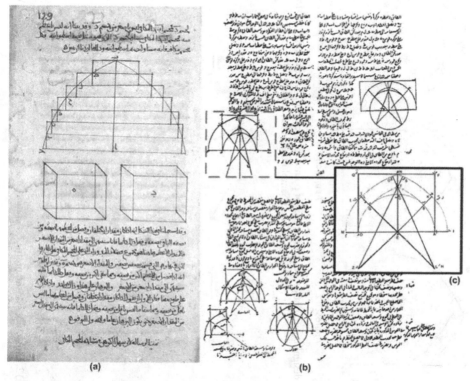

Fig. 10. Traditional approaches for designing dome's external shell: a) some geometrical shapes from Suhayl al-Quhi's book *Fi istihraci mesaha al-muhassama al-maqafi or Risala-i abu Sahl* [Suleymaniya library, Ayasofya- 4832]; b) a page of al-Kashi book, *IV Manuscripts* [Memarian 1988]; c) Dold-Samplonius re-drawing of the al-Kashi method [Dold-Samplonius 2000]

Dold-Samplonius has discussed the several aspects of the al-Kashi calculation principles. They include his good specific methods in approximating the surface areas and the volume of the shell forming of the *qubba* (the dome). She elaborated his five methods for drawing the "profile" of an arch from the al-Kashi' s *Key of Arithmetic* (fig. 11) [Hogendijk and Sabra 2003; Dold-Samplonius 1992].

– The first and second approaches addressed the design of a three-centered profile (centre points o, p, and q) according to the divisions of the circle into six parts (approach 1) and eight parts (fig.11, approach 2);

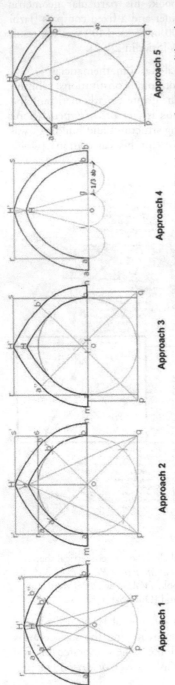

Fig. 11. Five approaches for designing and drawing different types of pointed arches from Kâshânî's *Key of Arithmetic*. After [Hogendijk and Sabra 2003; Dold-Samplonius 1992; 2000]

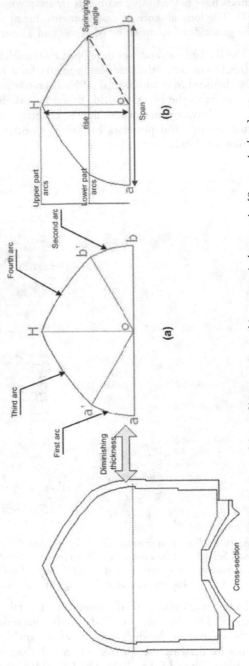

Fig. 12. Illustrations of the profile generation and its geometrical properties [Source: Authors]

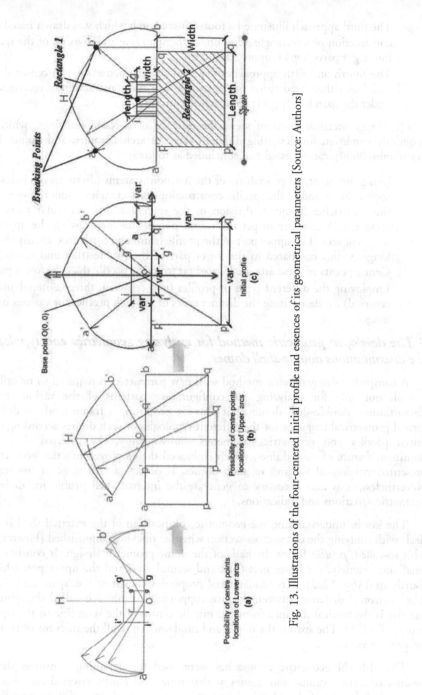

Fig. 13. Illustrations of the four-centered initial profile and essences of its geometrical parameters [Source: Authors]

- The third approach illustrated a four-centered arch which was drawn based on the construction of a rectangle *abpq* under the span line and division of the span line into eight parts (fig.11, approach 3);
- The fourth and fifth approaches depicted the geometries of two-centered arches based on either the division of the span line or the construction of rectangle *abpq* under the span line (fig.11, approaches 4 and 5).

On the geometrical point of view, the essences of al-Kashi drawings, which were frequently employed for designing different types of arches, vaults, and domes [Dold-Samplonius 2000], can be listed and concluded as follows:

- Using the variety of procedures of the division systems (divisions of circles in the approaches 1 and 2, the specific constructions of rectangles under the span line in the approaches 3 and 5, division of the span line into particular parts in the approach 4) in order to get the centre points and for drawing the appropriate shapes of arcs. The upper part of the profile (third and forth arcs) clearly showed a longer radius compared to the lower part of the profile (first and second arcs). Centre points of these arcs are located far from the profile than the lower part;
- Employing the different types of profiles (two-centered, three-centered and four-centered) for designating the distinct types of the arch profiles for various building usages.

3 The developed geometric method for analyzing geometries and typologies of the discontinuous double-shell domes

A comprehensive geometric method with new parameters is required to be utilized as a tool, not only for analyzing the configuration patterns of the various types of discontinuous double-shell domes, but also for proposing a frame work in defining a formal geometrical language for the different typologies of such domes according to their initial profiles and geometric parameters. Subsequently, the external shell as the dominant feature of such domes, which embraced the employed practice geometry and presented typological features of such domes, is object of this stage of investigation. Nevertheless, it is also necessary to generate the internal shell profile for deriving its geometric variations and indications.

The key in understanding the geometric composition of the external shell is mainly dealt with studying the dome cross-section when its thickness diminished [Huerta 2006]. This so-called '*profile*' forms the basis of the dome geometric design. It consists of four small arcs, namely, the lower part (first and second arcs) and the upper part (third and fourth arcs) (fig. 12a, b). Its fundamental properties consist of the 'span' and the 'rise'. The horizontal distance between the two supporting members is called the span whilst the rise is the vertical distance from the middle centre of the span line to the tip of the profile (fig.12b). The span is the origin and fundamental to all the rules for obtaining the proportion values.

The al-Kashi geometric approaches were used for designing common shapes of profiles of arches, vaults, and domes at that time, and hardly covered the majority of geometrical designs of the Islamic domes compositions with their different attributes at various periods. Nevertheless, the al-Kashi four-centered profile has the potential to be enhanced extensively through the expansion of the three geometrical parameters, including the location of the first and second arcs, the location of the third and forth arcs, and the positions of the breaking points.

On the other view, the frame work of the four-centered profile can be developed as 'a general initial shape' of the dome context. It is defined to provide the flexibility in its shape (or small arcs' curvatures) through the application of the following essential definitions (fig. 13):

1. *The lower arcs*: are the loci of the points whose centres are located on the two vertices of rectangle *ii'gg'* constructed above the span line. Values of its lengths and widths are gained based on the exact proportions of the span: $\dfrac{m_i}{n_i}$ *s*, (fig. 13a);

2. *The upper arcs*: are the loci of the points whose centres are always set on the two vertices of the rectangle *pp'qq'* constructed under the span line. Values of its lengths and widths are obtained based on the fractions of the span $\dfrac{m'_i}{n'_i}$ *s* (fig.13b); and

3. *The Breaking points*: are the couple of points *a'* and *b'* used for changing the profile curvatures through two considered options; firstly, it can occur by crossing the perpendicular lines *a"a'* and *b"b'* from the points *a"* and *b"* which are marked from the end points of the span line based on the fraction of span $\dfrac{m''}{n''}$ *s*. Secondly, the points *a'* and *b'* are gained from the certain values of the springing angles: α=25', 30' and 45'.

In fact, the general initial profile (fig. 13c) contains the whole essential geometric properties which are necessary in derivations of the typologies geometric prototypes and their geometric designs. To facilitate presenting and setting of its geometrical indications, a system is proposed $\{[R_1],(B),[R_2]\}$ = {Rectangular1, (Breaking point), Rectangular2} whilst all values are calculated from the middle point of the span, O (0, 0). Where *ab*=Span, and *i*=1, 2…5, then:

-   R_1: includes, respectively, values of a length and a width of the rectangle *ii'gg'* constructed above the span line. Two vertices of this rectangle are centre points of the lower part arcs with the variable values as:

$$\left[\begin{array}{l} ig = i'g' = \dfrac{m_1}{n_1} ab \\[2mm] ii' = gg' = \dfrac{m_2}{n_2} ab \end{array} \right].$$

When *ii'*=*gg'*=0, then the centre points are located on the span line.

-   B: shows two possible options of the breaking points either as the exact angular values ∠O=25°, 30° and 45°, or the coordinates values of points which are symmetrically positioned form the end points of span line as (*aa"*=*bb"*=$m_3/n_3 ab$,0);

-   R_2: describes, respectively, the values of a length and a width of rectangle *pqp'q'* constructed under the span line. Two vertices of this rectangle are the centre points of the upper arcs as follows:

$$\left[\begin{array}{l} p'q' = pq = \dfrac{m_4}{n_4}\,ab \\[2ex] qq' = pp' = \dfrac{m_5}{n_5}\,ab \end{array}\right].$$

Note that the string "var", which means 'variance' in the drawings, is a specific dimension or distance for having varied parameter on the specific direction. This function helps in the flexibility of the analysis and in defining the rules for generating common prototypes for the given typologies' profiles.

The given lengths in this system should divide by two for obtaining the vertices i, g, p, and q of the proposed rectangles. In fact, the obtained vertices are symmetrically calculated and positioned on the both sides of the vertical axis.

4 The morphological features of the discontinuous double-shell domes in the Middle East and Central Asia

4.1 Case studies

In order to define typologies of the discontinuous double-shell domes and to derive the associated geometric profiles, a sample of twenty one cases were investigated in the Middle East and Central Asia, including domes in Afghanistan, Iran, Azerbaijan, Turkey, Kazakhstan, and Uzbekistan. The architecture chronology of the eastern dome with their highlighted study periods is shown in fig. 14. The studied periods and the selected dynasties are particular causes for developing the dome configuration design that it never surpasses into the other periods later on. Fig. 15 exhibits twenty one assorted examples of the discontinuous double-shell domes according to their studied dynasties.

Fig. 14. Eastern dome chronology and the specific periods of the discontinuous double-shell domes

4.2 Common morphological features of the discontinuous double-shell domes

In general, the discontinuous double-shell domes expose logical designs, with dynamic articulations of distinct proportional components. They were extensively erected in the mausoleum buildings rather than whether the mosques or *madrasas* which are increasingly typical buildings of the late Islamic era.

As a result of the analytical analysis of the case studies, the common identical components of this type of the Islamic domes are morphologically addressed and listed as follows:

– *External shell:* this is what appeared from the outside of the dome buildings. It is the only architectural item which was conceptually found in synonymous during the several Islamic epochs. Its thickness is proportionally reduced from its base to the tip at either 25° or 30° angles;

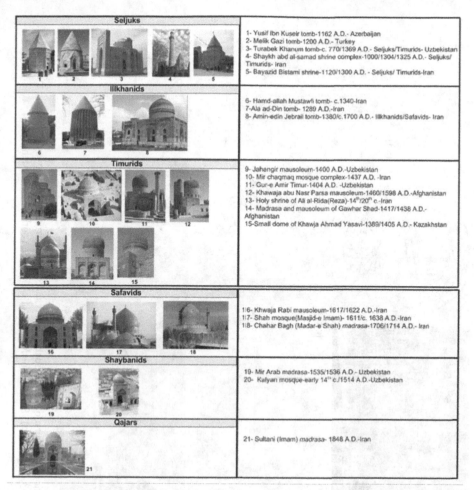

Fig. 15. Classification of selected samples based on the dynasties studied

- *Internal shell:* this covered the internal domical chamber and has a simple geometric formation compared to the external shell. In fact, it is necessary that its geometric shape fully conformed with the external shell for transferring forces from the upper components to the transition tier in order to stabilize the dome structure.

- *Drum:* Considerable thoughts and efforts were often given by the designers to make the building as high as possible using the tall drum. In fact, the discontinuous double-shell domes are considered as one of the highest samples of the eastern domes, with an average height of 30-35 meters from the ground. Its thickness must be sufficiently massive in order to transfer and neutralize the vertical thrust of the external shell to the lower items, especially, the internal shell;

- *Internal stiffeners with the wooden struts:* these are architecturally specific characteristics of such domes (except for the conical samples) called, 'the composition of the radial brick walls with the wooden struts'.

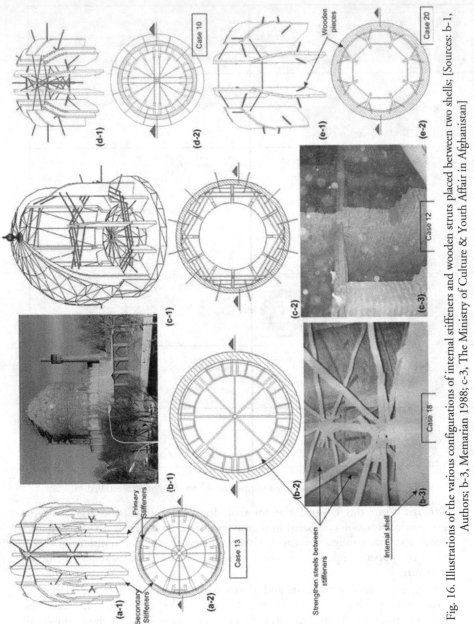

Fig. 16. Illustrations of the various configurations of internal stiffeners and wooden struts placed between two shells; [Sources: b-1, Authors; b-3, Memarian 1988; c-3, The Ministry of Culture & Youth Affair in Afghanistan]

Generally speaking, they were built in the space between the two shells for the main purpose of filling the empty spaces and to support the external shell. The radial walls are divided into two types of primary and secondary (smaller than primary ones, see fig. 16, a1-2). A different number of the radial walls observed were 4 (Sultan Bakht Aqa, first sample), 6, 8, 14, to 18. Their heights also varied from 8 to 15 meters. Their settings and sizes strongly affiliated with the size of the span. The walls thickness is also changed, but not less than 40 cm has been seen (fig. 16, c-3).

It is necessary to say that the arrangement of the internal stiffeners and their numbers have often been altered (fig. 16, c1-2)as a result of conservational interventions, such as replacement of the wooden struts with steel meshes or even removal of the wooden beams (fig. 16, b1, 2, and 3). Nevertheless, these removable wooden struts have basically set into the devised holes located on the thickness of the brick radial walls (stiffeners).

They are arranged compositionally based on the vernacular architecture agreements. For example, the vertical wooden posts were often seen in the Iranian samples (fig. 16, d1, 2). However, the pairs of the small wooden pieces, which are connected the radial walls to the drum's body, are the predominant features in the Uzbekistan, Turkmenistan, and Afghanistan (fig. 16, e1, 2).

5 The generative system for derivations of typologies and common geometries of the discontinuous double-shell domes

5.1 Samples analysis approach

Islamic domes, however, present a wide variety of sizes and types, but some geometric properties were repeatedly used in their composition designs. Nevertheless, no two samples are exactly the same. Visually, the analysis of the external shells of samples revealed three classifications of typologies including, conical, pointed, and bulbous which embrace the different geometrical properties and architectonic characteristics.

In primary stage, the computation process of the samples started with generation of both (external and internal) shells profiles as shown in stage 1 of fig. 17. In order to categorize the derived profiles of the external shells into the identified typologies, in what follows, the three shape-patterns of those typologies are mainly elaborated and schematically designed:

- Conical pattern: it is a triangle which is circumscribed by a rectangle;

- Pointed pattern: it is the feature whose lower arcs (the first and second arcs) are tangent to the two vertical lines passing from the end points of the span line; and

- Bulbous pattern: it is the prototype where the vertical lines intersected the lower arcs (the first and second arcs).

These patterns helped not only to distinguish between the pointed and bulbous profiles but to understand also some geometric properties of the deduced profiles constitutions which included location of the centre points of the lower arcs and proportions of rise to span of the conical samples.

In the second stage, after categorizing profiles in the exact typologies, their geometric parameters, proportions, angles, and breaking points locations were respectively deduced to identify the typical attributes of each typology.

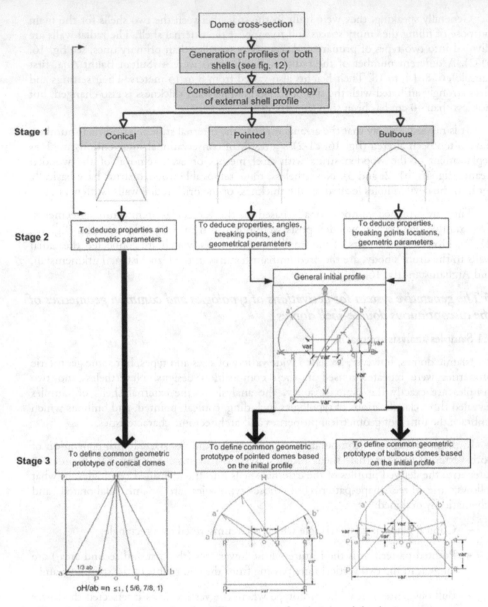

Fig. 17. A flow chart illustrating the different stages of analyzing and developing common geometric prototypes for the identified typologies of the discontinuous double-shell domes

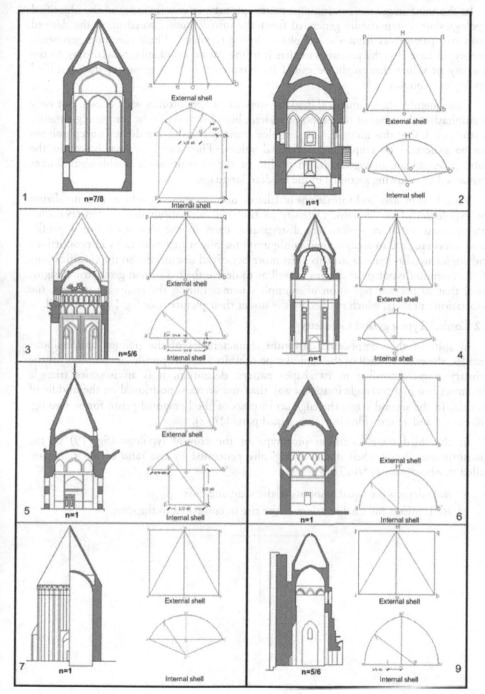

Fig. 18. Systematic geometric analysis of the conical typology samples. Cross-sections after:
[Hillenbrand 1994]; [Stierlin 2002]; [Pope 1971]; [Memarian 1988]; [The Iranian cultural
heritage organization documentation centre, (ICHO)]

In the third stage, the common geometric prototypes (see fig. 13c) of the identified typologies are systematically generated from the initial profile, based upon the derived geometric parameters from the samples of every typology. These common geometric prototypes have also the potential to alter into the variations of subsets according to the diversity of values that might be exited in parameters of each group of the identified typologies examples.

Subsequently, the geometrical drawing steps of sample profiles were drawn, not only to emphasize advantage of the proposed system, but also to verify the common geometric prototypes. Using this method, any complex compositions of the dome conceptualisms can be generated by employing additional values. They can be drawn based on the elaborations of an analogous drawn profile of the discontinuous double-shell domes (explained in following sections) in the modern language.

Variations in sizes and formations of Islamic domes expose a problem in generalizing the typologies of the designs. Diversity in the shell formations due to conservational interventions or other reasons are disregarded, even if, the developed initial profile parameters are able to analyze their configurations. Slight inconsistencies in proportions and angles are also ignored so as to assist more beneficial discussions on the classifications of the common typological designs as well as to derive their common geometric designs. Note that to prevent repetition of example's names during the analyzing process, the association numbers, which marked boldly under their pictures (see fig. 15), were used.

5.2 Conical Typology and Geometry

In spite of the complicated curvature characteristics of the pointed and bulbous profiles, the external shell of the conical type exhibits a simple profile constitution. In the primary stage, according to its shape- pattern definition, it is an isosceles triangle circumscribed in a rectangle in such a way that one vertex is positioned on the middle of its side. In the second stage, the edge ab in cases of the hexagonal prism forms (see fig. 18, cases 4 and 1) is divided into three equal parts (1/3 s).

In the third stage, common prototype of the conical typology (fig. 19) of the discontinuous double-shell domes is logically generated by the ratio of the rise/span, called n, where $n \leq \{1, 5/6, 7/8\}$:

- $n=1$: depicts the equal amounts of the span and rise;
- $n< 1$: means the smaller values of the rise in compared to the span.

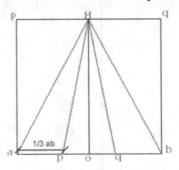

Fig. 19. Common geometric prototype of the conical typology. $\dfrac{oH}{ab} = n \leq 1, \left\{\dfrac{5}{6}, \dfrac{7}{8}, 1\right\}$

Depending on the design requirements and preferences, the forms of the internal shell showed variations, even more than the pointed and bulbous examples such as, semicircular (case 6), saucer (cases 2, 4, and 7), and several shapes of the pointed form with distinct geometrical properties including, the two-centered (cases 3, 5) and the four-centered (case 1) forms. In the pointed cases, the geometric parameters of profiles are systematically formulated based on the proposed system, which can facilitate their drawing steps, as follows:

- Case 1: $\left\{\begin{bmatrix} 1/3ab \\ 0 \end{bmatrix}, \angle 45, \begin{bmatrix} ab \\ ab \end{bmatrix}\right\}$;

- Case 3: $\left\{0,0, \begin{bmatrix} 1/2ab \\ 1/8ab \end{bmatrix}\right\}$;

- Case 5: $\left\{0,0, \begin{bmatrix} 1/2ab \\ 1/2ab \end{bmatrix}\right\}$;

- Case 9: $\left\{\begin{bmatrix} 2/6ab \\ 0 \end{bmatrix}, 0,0\right\}$.

The drawings of the conical external shell profiles are limited to the designation of triangles and rectangle, while the drawing steps of their internal shells with the pointed profiles fully conformed to the pointed typology that would be comprehensively elaborated in the next section.

No specific developments were discovered following the considerations of geometries and the architectonic compositions of the Seljuks mausoleums, except for the domical buildings which were partially reconstructed in the Timurids era.

5.3 Pointed Typology and Geometry

Historically, the pointed typology contained most number of the discontinuous double-shell domes in the Middle East and central Asia. In the primary stage, using the shape-pattern definition of this typology, if samples' lower arcs (primary and secondary) are tangent to the two vertical lines passing from the end of span, then the generated profile is categorized in this group.

To embrace such a property, the centre points of the lower arcs (first and second arcs) have to be set on the span line, meaning that $gg'=ii'=\frac{m_1}{n_1}s=0$. In addition, since the proportions of the breaking points ($\frac{m_3}{n_3}s$) on the span line do not show clear values, the exact angles (such as, 25°, 30°, and 45°) have thus been proposed for determining the breaking points.

In the second stage, as the results of the geometrical analysis of the examples, three subsets of the pointed profiles have been recognized and configured according to the number of their centre points (fig. 20):

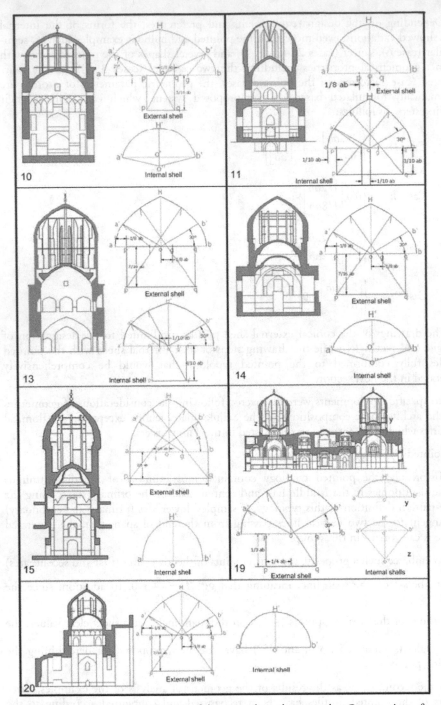

Fig. 20. Systematic geometric analysis of the pointed typology samples. Cross-sections after:
[Hillenbrand 1994]; [Stierlin 2002]; [Pope 1971]; [Memarian 1988]; [The Iranian cultural
heritage organization documentation centre, (ICHO)]; [The Ministry of culture & Youth Affair in
Afghanistan]; [Grangler 2004]

- Two-centered profile: The case 11 was analyzed based on the al-Kashi method and organized using the parameters of the new system:

$$\left\{ \begin{bmatrix} 2/8ab \\ 0 \end{bmatrix}, 0, 0 \right\}.$$

- Three-centered profile: It is only identified in the case 19 according to the following geometrical parameters:

$$\left\{ \begin{bmatrix} 0 \\ 0 \end{bmatrix}, \angle 30, \begin{bmatrix} 2/4ab \\ 1/3ab \end{bmatrix} \right\}.$$

- Four-centered profiles: The majority of the cases 10, 13, 14, 15, and 20 are categorized in this group based on the distinct geometrical indications:

- Case 10:
$$\left\{ \begin{bmatrix} 4/8ab \\ 0 \end{bmatrix}, \angle 25, \begin{bmatrix} 2/8ab \\ 5/16ab \end{bmatrix} \right\};$$

- Cases 13, 14, 20:
$$\left\{ \begin{bmatrix} 2/8ab \\ 0 \end{bmatrix}, \angle 30, \begin{bmatrix} 6/8ab \\ 7/16ab \end{bmatrix} \right\};$$

- and Case 15:
$$\left\{ \begin{bmatrix} 2/6ab \\ 0 \end{bmatrix}, \angle 45, \begin{bmatrix} ab \\ 2/6ab \end{bmatrix} \right\}.$$

In the third stage, the common geometric prototype of the pointed typology is systematically generated based on the geometrical indications through this system $\{[R_1], \angle B, [R_2]\}$, where its common geometrical parameters are presented as follows (see fig. 21):

$$\left\{ \begin{bmatrix} \frac{m_1}{n_1} \times ab \\ 0 \end{bmatrix}, \angle B, \begin{bmatrix} \frac{m_2}{n_2} \times ab \\ \frac{m_3}{n_3} \times ab \end{bmatrix} \right\}, \text{ ab=span}, \angle B = 25°, 30°, 45°, m_i \text{ is an integer digit and is}$$

greater than zero, ($m_i \leq n_i$); n_i= 3, 4, 6, 8, 16, and i=1, 2, 3.

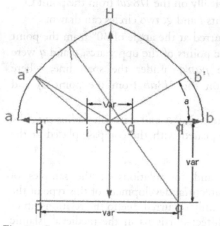

Fig. 21. Common geometric prototype of the pointed typology

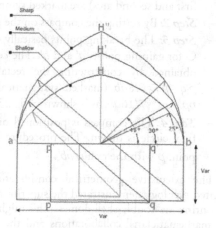

Fig. 22. Classification of the pointed typology based on the variety of rises

Therefore, based on the foregoing epistemological syntax of geometric variables, any kind of the pointed typologies of the Islamic discontinuous double-shell domes can be organized using these geometric indications.

Furthermore, according to the variance in heights of the rise, the external shells of this typology can also be categorized into shallow (see fig. 22, case 10), medium (cases 11, 13, 14, 19 and 20), and sharp (case 15). This property fully conformed to the scale of the rectangle $pp'qq'$ and the values of angle α.

The internal shell forms are recognized as semi-circular (cases 10, 14, and 20), semi-ellipse (cases 19y and 15), pointed (cases 11 and 13), and saucer forms (case 19z). On the other perspective, the presented analysis approach with its systematical presentation tool not only can utilize as the syntax elaboration of the geometric arrangements of the external shell of the discontinuous double-shell domes, but it also has the potential to compose any sort of the pointed features, such as the internal shell in given typologies as follows (see fig. 20):

- case 13: $\left\{ \begin{bmatrix} 2/10ab \\ 0 \end{bmatrix}, \angle 30, \begin{bmatrix} 4/10ab \\ 4/10ab \end{bmatrix} \right\}$;

- case 11: $\left\{ \begin{bmatrix} 2/10ab \\ 0 \end{bmatrix}, \angle 30, \begin{bmatrix} 8/10ab \\ 3/10ab \end{bmatrix} \right\}$.

To verify the common geometric prototype, the four-centered profile with its geometric parameters for instance:

$$\left\{ \begin{bmatrix} 2/8ab \\ 0 \end{bmatrix}, \angle 30, \begin{bmatrix} 3/4ab \\ 7/16ab \end{bmatrix} \right\},$$

which is considered as most common geometric attributes of the pointed samples, has been selected. Note that the same method can also be adapted to draw any sort of geometric prototype of the pointed profile. The construction of the sample is thus (fig. 23):

- **Step 1:** The baseline ab as span is constructed and divided into 8 equal parts. The point O is set on the $1/2ab$. The points i and g (named the centre points of the first and second arcs) are marked symmetrically on the $1/8\ ab$ from the point O.
- **Step 2:** By setting the compass on the points i and g, two circles can drawn.
- **Step 3:** The breaking points basically occurred at the angle of 30° from the point O for gaining points a' and b'. The centre points of the upper arcs, p and q were obtained by constructing the rectangle $pqp'q'$ under the span line, where $p'q'=pq=6/8ab$ (marked systematically on the $3/8ab$ from the point o) and $pp'=qq'=7/16ab$, as is shown in fig.23.
- **Step 4:** The compass is positioned on the point q and with the radius qa', the final circle is drawn. The procedure is repeated with the compass placed at the point p with the radius pb'.

The exhaustive geometrical considerations and formulations of the samples of pointed typology demonstrated the superior architectural development of this type of the discontinuous double-shell domes. This might mainly occurred due to the relationship of the mathematicians' collaborations and the architect's artisans in the medieval Islamic era.

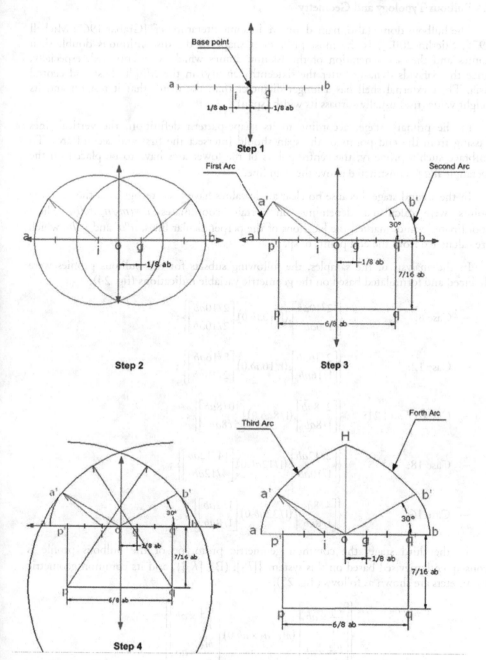

Fig. 23. Illustrations of geometrical drawing steps of the pointed profile typology

5.4 Bulbous Typology and Geometry

The bulbous dome (also, named 'onion' in some literature; cf. [Grabar 1963; Michell 1978; Stierlin 2002]) is the most prominent shape of the discontinuous double-shell domes and the last generation of the Islamic domes which were appeared, especially, since the Safavids dynasty (after the sixteenth century) in the Middle East and central Asia. Their external shell has a longer diameter than the drum that it rest on and its height value (rise) usually surpass its width (span).

In the primary stage, according to its shape-pattern definition, the vertical lines passing from the end points of the span should intersect the first and second arcs. To embrace such a property, the centre points of the lower arcs have to be placed on the rectangle $ii'gg'$, constructed above the span line.

In the second stage, because no clear angle values have been recognized, the breaking points were calculated depending on certain coordinates $B=(m_3/n_3ab,0)$. These coordinates' values marked the locations of the perpendicular lines, $a''a'$ and $b''b'$, which are calculated from the end points of span.

In the analysis of the samples, the following subsets for the bulbous profiles were deduced and formulated based on the geometric variable indications (fig. 24):

- Case 8: $\left\{ \begin{bmatrix} 2/10ab \\ 1/6ab \end{bmatrix}, (1/10ab,0), \begin{bmatrix} 2/10ab \\ 2/10ab \end{bmatrix} \right\};$

- Case 12: $\left\{ \begin{bmatrix} 2/16ab \\ 1/16ab \end{bmatrix}, (1/16ab,0), \begin{bmatrix} 8/16ab \\ 2/16ab \end{bmatrix} \right\};$

- Cases 17 and 21: $\left\{ \begin{bmatrix} 2/8ab \\ 1/8ab \end{bmatrix}, (1/8ab,0), \begin{bmatrix} 10/8ab \\ 5/8ab \end{bmatrix} \right\};$

- Case 18: $\left\{ \begin{bmatrix} 2/12ab \\ 1/10ab \end{bmatrix}, (1/12ab,0), \begin{bmatrix} 14/12ab \\ 4/12ab \end{bmatrix} \right\};$

- Case 16: $\left\{ \begin{bmatrix} 4/8ab \\ 1/8ab \end{bmatrix}, (1/32ab,0), \begin{bmatrix} 1/2ab \\ 1/8ab \end{bmatrix} \right\}.$

In the third stage, the common geometric prototype of the bulbous profile is consequently devised based on this system $\{[R_1], (B), [R_2]\}$, and its common geometric parameters are shown as follows (fig. 25):

$$\left\{ \begin{bmatrix} \dfrac{m_1}{n_1} \times ab \\ \dfrac{m_2}{n_2} \times ab \end{bmatrix}, (m_3/n_3 \times ab,0), \begin{bmatrix} \dfrac{m_4}{n_4} \times ab \\ \dfrac{m_5}{n_5} \times ab \end{bmatrix} \right\},$$

where m_i is an integer digit (it is possible that $m_4 > n_4$) and is greater than zero, n_i=6,8,10,12,16,32 (only in n_3), ab= span, and i=1, 2...5.

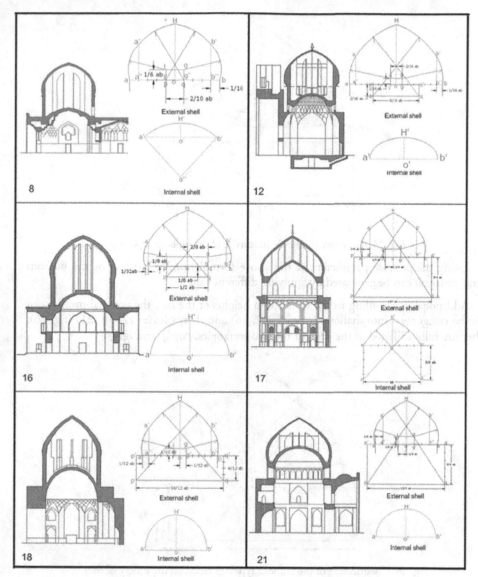

Fig. 24. Systematic geometric analysis of the bulbous typology samples. Cross-sections after:
[Stierlin 2002]; [Memarian 1988]; [The Iranian cultural heritage organization documentation
centre, (ICHO)]; [The Ministry of culture & Youth Affair in Afghanistan]

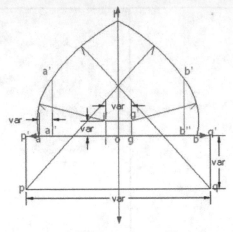

Fig. 25. Common geometric prototype of the bulbous typology

Using the premise parameters of the above formulation, any subset of the bulbous conceptualism can be generated by applying different values.

Additionally, according to the variance in heights of the rise, the external profiles can also be categorized into shallows (cases 8, 12, 16) and sharps (cases 17, 18, 21) (fig. 26). They are fully affiliated to the scale of the two rectangles, *pqp'q'* and *igi'g'*.

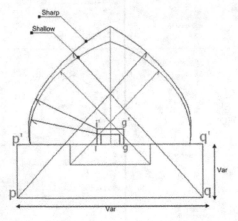

Fig. 26. Classification of the bulbous typology based on the variety of rises

Architecturally, the internal shell formations have been found that resemble the pointed typology in most cases, such as the saucer (case 8), semi-elliptical (cases 12 and 18), and semi-circular (cases 16 and 21) forms. However, bulbous geometric attributes (case 17) did not show the developing configuration compared to either the conical or pointed typologies. The geometric parameters of this sample can be set out as follows:

$$\left\{ 0, 0, \begin{bmatrix} ab \\ 5/8ab \end{bmatrix} \right\}.$$

The bulbous profile of the case 12 with the geometric indications

$$\left\{ \begin{bmatrix} 2/16ab \\ 1/16ab \end{bmatrix}, (1/16ab, 0), \begin{bmatrix} 8/16ab \\ 2/16ab \end{bmatrix} \right\}$$

was selected for the verification of common geometric prototype of bulbous. It is interesting to say that the same approach can be used for drawing any kind of the bulbous profiles. The construction of the bulbous profile is thus (fig. 27):

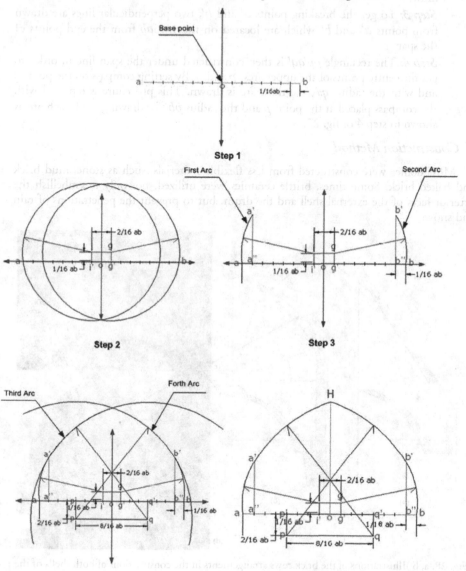

Fig. 27. Illustrations of geometrical drawing steps of the bulbous profile typology

- *Step 1:* The base line of the span, namely, *ab* is constructed and then divided into 16 equal parts. The point O is set on the 1/2*ab*.
- *Step 2:* The rectangle *igi'g'* is then marked out symmetrically on the 2/16 *ab* (length) and 1/16 *ab* (width) from the base point O, constructed above the span line. By setting the compass on the points *i* and *g*, two circles are respectively drawn.
- *Step 3:* To get the breaking points *a'* and *b'*, two perpendicular lines are drawn from points *a"* and *b"* which are located on the 1/16*ab* from the end points of the span.
- *Step 4:* The rectangle *pp'qq'* is then constructed under the span line in order to get the centre points of the upper arcs, *p* and *q*. By setting compass on the point *q* and with the radius *qa'*, the third arc is drawn. This procedure is repeated with the compass placed at the point *p* and the radius *pb'* for drawing the fourth arc as shown in step 4 of fig. 27.

6 Construction Method

Most domes were constructed from less flexible materials, such as stone, mud brick and baked brick. Some times, brittle ceramics were utilized, not only to embellish the exterior faces of the external shell and the drum, but to prevent the penetrations of rain and snow.

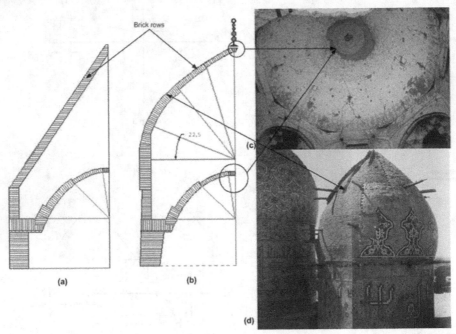

Fig. 28. a, b)Illustrations of the brick rows arrangements in the constructions of both shells of the discontinuous double-shell domes; c) Sayyidun tomb, Ilkhanids, *Abarquh*, Iran ; d) Domes of Darb-i Imam shrine, Safavids, Isfahan, Iran. [Photos: www.archnet.org, photographers: Sheila Blair and Jonathan Bloom 1984]

The common method for erecting the discontinuous double-shell dome involves the building of half of the internal shell within the main roof. The drum and the stiffeners were then built together on the lower components. The last task is to close the external shell and to set the wooden struts in the devised holes between the radial walls (fig. 28d). Nevertheless, it was impossible to continue using the same construction method near the apse of both the shells where the empty oculi were remained. Therefore, the rows of bricks were arranged vertically (fig. 28c) at these parts. The oculus of the internal shell is the last task of construction. The construction techniques for arranging the brick layers varied depending on the eastern domical types.

In bulbous and pointed typologies, the directions of the bricks' rows are almost always perpendicular to the generated curve of the dome surface (fig. 28b) in both the external and internal shells. The thicknesses of both shells were also gradually reduced at whether 22.5 or 25 angles for decreasing weights of the shells (except for the external shell of conical samples). In the conical external shells, bricks rows are arranged with their horizontal directions in such a way that the bricks in the upper row are precisely laid on the half part of those in the lower row (fig. 28a). Similarly, the construction techniques of internal shell of the conical typology fully conformed to that of curvature's typologies.

Conclusion

The discontinuous double-shell dome is defined as the dome whose shells have considerable distances. The discontinuous double-shell domes presented the high level of design and configuration in the Islamic dome architecture in the Middle East and Central Asia. They also included the majority of typologies of the Islamic domes from the early Islamic era through to the late Islamic epoch.

The present research involved a new frame work that, not only identified systematically morphological features and typologies variations of the discontinuous double-shell domes, but also offered the analytical understanding of their geometrical prototypes and related parameters which can be appeared in the designs of the Islamic dome architecture during various eras.

Based on the essences of the al-Kashi geometric approach, a theoretical frame work was geometrically developed in order to exhibit the geometric principles of compositions of the discontinuous double-shell domes. By using the developed initial profile with its defined parameters, the formal language for the discontinuous double-shell domes has systematically been specified. It included: redefinitions of identified typologies, classifications of the typologies subsets according to their heights variations, derivations of internal shell forms and their geometric indications. Twenty-one samples of the discontinuous double-shell domes, which were built in the Middle East and central Asian countries, were subjects of this analysis.

Four main components morphologically recognized such as, the external shell, the internal shell, the drum, and the internal stiffeners. Three main typologies of such domes have geometrically derived and redefined including, conical, pointed, and bulbous. Three subsets for the pointed and bulbous typologies have commonly derived based on their external shells heights such as, shallow, medium, and sharp.

The biggest challenge was the creation of flexibility in configuring the initial profile, especially, in locations of its breaking points that allowed the possibilities of covering various dome designs. The results geometrically included two optional characters for the breaking points.

The generated initial profile with its developed geometrical parameters, and the proposed system, has the potential of offering a unique computational method for the design analysis and geometric derivations of the Islamic domes, such as the Mongolians, one shell domes, etc. Simultaneously, the results of this systematic geometrical analysis could be recognized by using whether shape grammar or genetic algorithm, not only to generate any kind of the Islamic domical configuration, but also to develop a more advanced archive and retrieval system for similar data.

References

ASKAROV, S. 2003. Splendour of Bukhara. *Archive of San'at magazine* **4**. (http://www.sanat.orexca.com/eng/4-03/history_art2.shtml, accessed 19 Sep 2008)

CRESWELL, K. A. C. 1958. *A short account of the Early Muslim architecture II*. Oxford: Clarendon Press.

DOLD-SAMPLONIUS, Y. 1992. The 15th century Timurid mathematician Ghiyath al-Din Jamshid al-Kashi and his computation of the Qubba. *Amphora : Festschrift for Hans Wussing on the occasion of his 65th birthday*. S. S. Demidov et al. (eds.), Basel, Boston, Berlin, p. 171-181.

———. 2000. Calculation of Arches and Domes in 15th Century Samarkand. *Nexus Network Journal* **2**, 3: 45-55.

FARSHAD, M. 1977. On the shape of Momentless Tensionless Masonry Domes. *Journal of Building and Environment* (**12**): 81-85.

GRABAR, O. 1963. The Islamic dome: Some considerations. *The journal of the society of Architectural Historians* **22**, 4: 191-198.

———. 2006. *Islamic Art and Beyond*. United Kingdom:Variorum Press.

GRANGLER, A. 2004. *Bukhara: The Eastern dome of Islam*. Stuttgart: Axel Menges Press.

HEJAZI, M. M. 1997. *Historical Buildings of Iran: their Architecture and Structure*. Southampton: Computational Mechanics Publications.

———. 2003. Seismic Vulnerability of Iranian historical Domes. *Journal of Earthquake Resistant Engineering Structures* **4**: 157-165.

HUERTA, S. 2006. Galileo was Wrong: The geometrical Design of Masonry Arches. *Nexus Network Journal* **8 (2)**. 25-52

HILLENBRAND, R. 1994. *Islamic architecture: form, function, and meaning*. New York: Columbia University Press.

———. 1999. The Ilkhanids and Timurids. In *Islamic Art and Architecture*. London: Thames and Hudson. (p. 196-202)

HOAG, J. D. 2004. *History of World Architecture: Islamic Architecture*. Milan: Phaidon Press.

HOGENDIJK, J. P. and SABRA, A. I. (eds). 2003. *The enterprise of the Science in Islam: New perspectives*. Cambridge: MIT Press.

IRFAN, H. 2002. Sacred Geometry of Islamic Mosque. *Islamonline- News Section*. (http://www.islamonline.net/English/Science/2002/07/article02.shtml, accessed 5 Sept 2007)

JAZBI, S. A. (ed). 1997. *Applied Geometry: By Abolvafa mohammad ibn mohammad al-Buzjani* (in Persian and Arabic).Tehran: Soroush Press, ISBN 964 435 201 7.

KATZ, V. J. (ed). 2007. *The Mathematics of Egypt, Mesopotamia, China, India, and Islam: A source Book*. New Jersey: Princeton University Press.

KENNEDY, E.S. 1960. A Letter of Jamshid al-Kashi to His Father: Scientific Research at a Fifteenth Century Court. *Orientalia, Nova Series* **29**, 191-213.

MAINSTONE, R. J. 2001. Vaults, domes, and curved membrances. In *Developments on Structural Form*. London: Architectural Press. (p. 124)

MEINECKE, M. 1985. Mamluk Architecture, Regional Architectural Traditions: Evolution and Interrelations. *Damas-zener Mitteilungen* **2**: 163-175.

MEMARIAN, G. Hossein. 1988. *Statics of Arched Structures* (*Nîyâresh-e Sâzehâye Tâghî*), Vol. 1. Tehran: Iran University of Science and Technology Press. (In Persian)

MICHELL, G. (ed.). 1978. *Architecture of the Islamic world: Its History and Social Meaning*. New York: Thames and Hudson.

O'CONNER J. J. and ROBERTSON E. F. 1999. Abu Sahl Waijan ibn Rustam al-Quhi. In *School of Mathematics and Statistics, University of St Andrews, Scotland.* (http://www-history.mcs.st-andrews.ac.uk/Biographies/Al-Quhi.html, accessed 12 April 2009).

O'KANE, B. 1998. Dome in Iranian Architecture. In *Iranian Art and Architecture.* (http://www.cais-soas.com/CAIS/Architecture/dome.htm, accessed 22 June 2008).

ÖZDURAL, A. 1995. Omar Khayyam, Mathematicians, and "Conversazioni" with Artisans. *Journal of the Society of Architectural Historians* **54**, 1: 54-71.

———. 2000. Mathematics and Arts: Connections between Theory and Practice in the medieval Islamic world. *The Journal of Historia Mathematica* **27**, 2:171-201.

POPE, A.U. 1971. *Introducing Persian Architecture.* London: Oxford University Press.

———. 1976. Introducing Persian Architecture. In *Survey of Persian Art*, A. U. Pope and P. Ackerman (eds). Tehran: Soroush Press.

SAUD, R. 2003. Muslim Architecture under Seljuk Patronage (1038-1327). *Foundation for Science, Technology, and Civilization (FSTC).* (http://www.muslimheritage.com/features/default.cfm?ArticleID=347, accessed 3 Feb 2009).

SEHERR-THOSS, S. P. 1968. *Design and Color in Islamic Architecture: Afghanistan, Iran, Turkey.* Washington, D.C.: Smithsonian Institution Press.

SMITH, E. B. 1971. *The dome: A study in the History of Ideas.* New Jersey: Princeton Press.

STIERLIN, H. 2002. Islamic art and Architecture: From Isfahan to the Taj Mahal. London: Thames and Hudson.

About the authors

Maryam Ashkan was born in Iran in 1978, and graduated with the M. Arch degree in 2003 from Islamic Azad University/Qazvin, Dept. of Architecture and Urban Design. She started her career as conservator architect in Jameh Kavir Construction Company (2003-2006). Since July 2006, she has been a Ph.D candidate in the University of Malaya, Kuala Lumpur, in the Faculty of Built Environment specializing in historic Eastern domes in the Middle East and Central Asia.

Associate Professor Dr. Yahaya Ahmad received his Bachelor of Arts in Architecture (1986), Master of Construction Management (1987) and Master of Architecture (1988) from Washington University, USA; and his Ph.D in Conservation Management from the University of Liverpool, UK (2004). He has published many academic papers on conservation and is directly involved in many conservation projects. He was involved in the drafting of the National Heritage Act (2005) Malaysia, headed the nomination team to prepare the nomination dossier for Melaka and George Town for listing on the World Heritage List, and headed expert teams in the re-construction of Melaka fort. He was seconded to the Department of National Heritage as Deputy Commissioner of Heritage Malaysia 2007-2009, and was elected as ICCROM Council Member 2007-2011.

Giulio Magli

Dipartimento di Matematica
Politecnico di Milano
P.le Leonardo da Vinci 32
20133 Milano ITALY
Giulio.Magli@polimi.it

Keywords: Machu Picchu, Inca
architecture, Inca astronomy

Research

At the Other End of the Sun's Path: A New Interpretation of Machu Picchu

Abstract. The Inca citadel of Machu Picchu is usually interpreted as a "royal estate" of the Inca ruler Pachacuti. This idea is challenged here by a critical reappraisal of existing sources and a re-analysis of existing evidence. It is shown that such evidence actually point at a quite different interpretation suggested, on one hand, by several clues coming from the urban layout (the interior arrangement of the town, the ancient access ways, the position with respect to the landscape and the cycles of the celestial bodies in Inca times), and, on the other hand, by a comparison with known information about the Inca pilgrimage center on the Island of the Sun of Lake Titicaca. Altogether, these clues lead us to propose that Machu Picchu was intentionally planned and built as a pilgrimage center connected with the Inca "cosmovision".

1 Introduction

This paper analyzes one of the most beautiful and enigmatic achievements ever accomplished anywhere in the world by architecture. It is an ancient Andean town whose original name is unknown; it is famous with the name *Machu Picchu*. Although it may seem strange at a first glance for such a renowned archaological site, the reason why the town was built, the date at which it was built, the ruler who ordered its construction, the reason why it was abandoned, in a word, the *interpretation* of this place, are unknown. For reasons we do not know, Machu Picchu was abandoned and forgotten; it was brought again to the attention of the world only with the famous expedition of Hiram Bingham in 1911 (see [Bingham 1952] or [Salazar-Burger 2004] for an up-to-date account). Immediately after its "re-discovery" the site was enveloped by a halo of mystery. Bingham himself thought it to be the "lost capital" of the last Inca reign, Vilcabamba, an interpretation that we know to be untenable today; various errors and misunderstandings further contributed to the confusion, such as, for instance, an exaggeratedly high estimate of the percentage of female bones found in the burials, which led to the hypothesis that Machu Picchu may have been a sanctuary inhabited by Inca's "Virgins of the Sun". Today, this as well as other, even more unsound theories have been canceled by modern archaeological research (for instance, the true excess of female with respect to male bones is around 1.46 to 1). Modern research also helped, for instance, to clarify the day-to-day life of the inhabitants (see [Burger 2004] and references therein). However, contemporary with such important developments, nothing has really come out that helps to explain why and when the Incas built the town (or at least, this is the opinion of the present author, to be substantiated in what follows).

Living without interpretative schemes is extremely difficult in any science (e.g., physics) and archaeology is no exception. Therefore, archaeologists have adopted a scheme, a sort of *dogma*, about the true meaning of the town: the idea that Machu Picchu is to be identified as one of the "royal estates" of the Inca Pachacuti [Rowe 1990].

Nexus Network Journal 12 (2010) 321–341   NEXUS NETWORK JOURNAL – VOL.12, No. 2, 2010 **321**
DOI 10.1007/s00004-010-0028-2; *published online* 4 May 2010
© 2010 Kim Williams Books, Turin

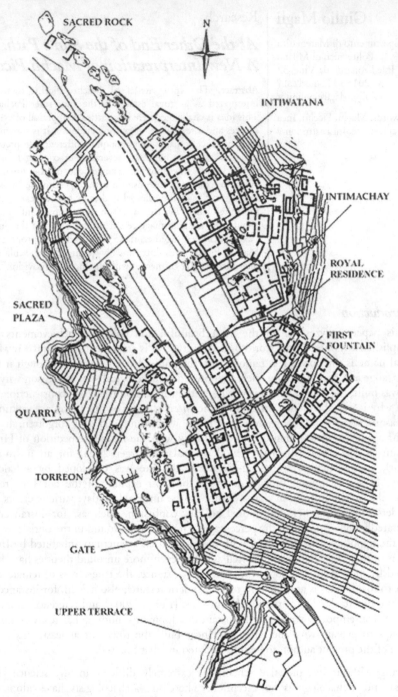

SACRED ROCK

N

INTIWATANA

INTIMACHAY

ROYAL
RESIDENCE

SACRED
PLAZA

FIRST
FOUNTAIN

QUARRY

TORREON

GATE

UPPER TERRACE

Fig. 1a. Map of Machu Picchu highlighting the sites discussed in the text (north on top)

Fig. 1b. Machu Picchu. Photograph by the author

What is today customarily called an Inca "royal estate" was a land property, nominally owned by the king and managed by his family clan. A royal estate was typically composed of agricultural lands and "palaces" intended to serve as residences for the ruler and the *elite*. Such places were used for amusement (such as hunting) and perhaps also for dealing with state affairs. A good example of royal estates is Chincero, property of Topa Inca, described in detail by the chronicler Betanzos as a property "where to go for recreation" and thoroughly analyzed by Niles [1999]. Other important Inca sites have been interpreted as royal estates as well, in particular Pisac and Ollantaytambo. To the best of the present author's knowledge, however, there is no textual evidence whatsoever showing that Pisac was a royal estate (Pisac is never mentioned in the Spanish chronicles). More convincing is the case of Ollantaytambo [Protzen 1993]. This place is in fact associated with the ruler Pachacuti in some Spanish accounts and, in particular, Sarmiento de Gamboa says that the king "took as his own the valley...where he erected some magnificent buildings". As far as Machu Picchu is concerned, it is easy to get the impression from most of the scholarly work that there exist firm textual evidence that this town too was one of Pachacuti's royal estates. However, as discussed in full detail here in Appendix 1, this is *not* the case. It is the aim of the present paper to propose a completely different theory regarding the conception and construction of the town. To "make room" for such an interpretation I am going to use a scientific instrument which is typical of a physicist's way of thinking: Occam's razor. According to this principle, what is unproved and unnecessary not only can, but should be cut away from any scientific approach to a problem. I thus attempt to show (Appendix 1) that actually *there is no proof whatsoever*, either textual or archaeological, that Machu Picchu was built as a private estate for Pachacuti (or, for that matter, for any other Inca ruler). The interpretation as a royal estate therefore is both unproved and unnecessary: Occam's razor allows us to cut it away. Further, although I do not exclude that the site may have had *secondary* functions, the "multi-functional" interpretation adopted by some scholars, who view Machu Picchu as a royal estate but also as a sacred and perhaps administrative center as well (see e.g. [Reinhard 2007], [Malville and Ziegler 2007]) is also refuted here. The reason again is that this point of view does not help to explain those characteristics of the site (such the urban layout, for instance) which render Machu Picchu *unique* among the Inca towns, and those other characteristics (such as the clear directionality in access and fruition) which render Machu Picchu *almost* unique, the sole possible comparison being a place whose main function had nothing to do with royal estates or multi-functional centers: the sanctuary on the Island of the Sun (discussed in § 4).

All in all, as a working hypothesis in what follows I will neglect any pre-existing interpretation. This means trying to explore the problem of the meaning of Machu Picchu starting from, and only from, the very beginning: the town itself.

2 A "non-standard" description of Machu Picchu

Machu Picchu lies at 2400 m. above sea level, built like a condor's nest between the two twin peaks (Huayna Picchu to the north and Machu Picchu to the south) which form a sort of peninsula, surrounded on three sides by the gorges of the Urubamba River some 80 km. northwest of Cusco, the capital of the Inca empire. It would be of course out of the scope of the present paper to give a full description of the site; however, we need a clear idea of the urban layout and the relationship of the town with the landscape. Here we immediately encounter a curious problem. Indeed, since the city is stretched in an approximate southeast-northwest direction (in conformance to the general direction of the Machu Picchu-Huayna Picchu ridge), a space-saving device on the printed page usually results in – starting from the very first Bingham's plan –north appearing on the ut lower right. However, this way of mapping the town completely changes the correct perspective in which the place is – and, most important, *was* – actually *visited* by a newcomer. Further, this unusual orientation makes it difficult to understand at a glance the orientation of the buildings with respect to the path of the sun and of the stars, as well as the relationships of the layout of the town with the cardinal points (and the mountains associated to them). Therefore, although it may seem a somewhat trivial point, I consider it *fundamental* for a correct interpretation of the site to refer to a map that shows north on top (to this aim I have rotated one of the original maps made by Bingham; see fig. 1).

Further, to understand the layout of Machu Picchu, it is important to keep in mind that the complete urban plan was conceived, planned, founded and built from scratch, following the precise will of the planner on the basis of a global, unitary project laid out after an accurate and complete survey of the area and, in particular, of the natural rock outcrops which were scattered around. Many such rocks were leveled; others, however, were "interpreted" in an artistic way, in accordance with that special feeling of the Inca artists for the stone and, in particular, for the shapes of stones that "replicate" natural elements. Huge boulders and existing caves were therefore wonderfully integrated into the project. Each time the planner felt it necessary, the terrain was leveled with the use of a sophisticated technique of superimposed foundation layers. It is thought that as much as 60% of the work involved in the construction of the town is buried in its basements [Wright and Valencia Zagarra 2001]. Finally, it is important to notice that the town is lacking in water sources, and therefore a careful project was needed to construct an aqueduct which brings water from a spring located higher on the north flank of Machu Picchu mountain.

The town was abandoned when this huge building program was near to completion, so that some elements – such as for instance the area of the Temple of the Three Windows – were left unfinished, as it was, probably, another magnificent architectural project of the Incas, namely the Sacsahuaman "fortress" in Cusco [Protzen 2004]. It is of course difficult to estimate the length of time employed to bring the site to the present almost-finished state (to the best of my knowledge, nobody has ever tried to figure out how long this took). In any case, it is hard to believe that from the beginning of the planning to the state visible today – with the inclusion of hundreds of agricultural terraces – it could have taken less than, say, a number of decades (decidedly a long time to wait for a private estate to be ready).

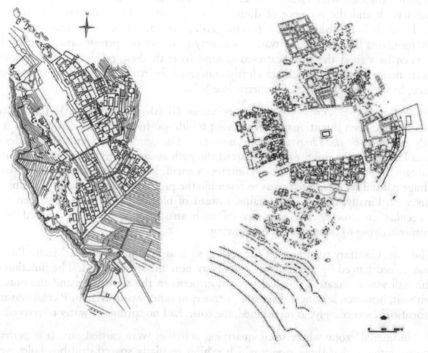

Fig. 2. Comparison between maps of Machu Picchu and Huanuco Pampa (north on top in both, but maps are not at the same scale)

The rigorous, unitary project inspiring the construction of the town remained fully unaltered after the conquest, contrary to what happened to the other Inca settlements. There is, however, a notable exception: the provincial site called Huanuco Pampa [Morris & Thompson 1985]. This town was created by the Incas as an administrative center, and as such it has nothing to do with Machu Picchu from a functional point of view. However, it is very useful as a term of comparison in order to gain a better understanding of the inspiring principles of Inca town planning [Hyslop 1990]. Huanuco was founded in the central highlands of Chinchaysuyu, at 3700 m. above sea level. The Spaniards rapidly gave up the idea of living at such an unfriendly altitude, and therefore the place was soon abandoned (fig. 2, right). It exhibits four main "quarters" corresponding roughly to the cardinal directions and surrounding an enormous *plaza*, which is empty except for a rectangular building probably used for ceremonial purposes. The eastern sector of the town – connected to the center by a spectacular east-west alignment of double-jamb doorways – is the sole part built in fine Inca stonework and was certainly devoted to the rulers and to ritual activities.

If we compare the two plans, we immediately see that also Machu Picchu was apparently conceived in a pretty similar, quadripartite way (fig. 2, left).[1] The division of the town into sectors which will be used here (slightly different from the one usually adopted) is given in fig. 3; in it we can recognize a *key* difference with Huanuco Pampa: blocks of buildings at Machu Picchu are present only in two sectors (I and II). Most of the buildings in these two sectors can be understood through stylistic analyses of Inca architecture [Gasparini and Margolies 1980; Kendall 1985; Niles 2004]. Indeed most of the *Kanchas* (blocks) were clearly conceived as residences of the elite, as shown by the fine stonework and the presence of double-jamb doorways. In particular, the block of Sector I, which is furnished with a private garden and direct access to the first of the gravity fountains (i.e., the purest water), was very probably the private apartment of the ruler when he visited the site. Scattered around in both these residential quarters are, however, many buildings which are clearly conceived for ritual purposes, as shown, for instance, by their astronomical alignments (see § 3).

Proceeding anti-clockwise we encounter Sector III (the northernmost). This Sector appears to have been intentionally left without buildings. In a sense we can say that it is actually *occupied by the Huayna Picchu mountain*. The central plaza ends at the sides of Huayna Picchu; to the right, near the start of the path ascending to the summit, a visitor encounters the so-called *Sacred Rock* complex, a small, leveled plaza closed on the west by a huge natural rock sculpted so as to resemble the profile of Mount Yanatin, visible at the distance. Finally, Sector IV contains, instead of blocks of buildings, a sequence of very peculiar structures. The sequence of such structures as they are viewed by a newcomer arriving to the town is the following:

1) The gate. Contrary to almost all Inca cities, such as Huanuco Pampa, Machu Picchu was indeed fenced by a wall, with a doorway near the western end. The function of the wall was to create a physical separation between the settlement and the outside without, however, serving a defensive purpose; in other words Machu Picchu was not "fortified" (accordingly, as mentioned, the town had no springs or water reservoirs).

2) A "disordered" zone where stone-quarrying activities were carried out. It is perhaps worth noticing that those stones which exhibit regularly spaced drill-bits holes were *not* worked by the Incas but are the results of modern "archaeological" experiments; actually this method – typical of the Romans – was not used by the Incas. They instead used hard hammer-stones and took advantage of natural fissures of the rocks. Inca methods are clearly visible in many other points of the same area.

3) The so-called Sacred Plaza, a small space open to the western horizon and closed on the other three sides. To the east, in particular, one finds the so-called Temple of the Three Windows, actually a three-sided building. The windows, a spectacular feat of engineering composed of huge, perfectly dressed polygonal blocks, are located on the east wall facing the central plaza.

4) The so-called Intihuatana, a steep terraced pyramid on the summit of which lies a carved stone of white granite.

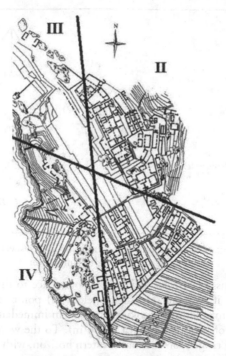

Fig. 3. The proposed "quadripartition" of Machu Picchu; see text for details

3 Machu Picchu and the landscape

Machu Picchu is wonderfully integrated into the landscape; this is so in a quite "global" sense, including not only the earthly landscape, but also the sky at the time of construction. The word "landscape" is therefore meant here in the broad sense which is commonly used today in archaeoastronomy (see e.g. [Magli 2009]).

To describe the relationship between the town and the landscape, it is worth starting from the analysis given by Reinhard [2007] of the position of the town with respect to the surrounding mountains. Indeed, as is well known, the mountains were a fundamental part of the Inca's *huacas* (shrines).

To visualize the position of Machu Picchu we can draw a cardinally oriented cross with the town at the center, and then follow the four cardinal directions (fig. 4). In this way we meet four important peaks of the region. First of all, north is identified by the "proprietary" mountain Huayna Picchu, which is, in a sense, a part of the town itself. The summit (2720 m) was – and is – accessed via a spectacular path, essentially a steep flight of steps partly carved into the rock which circles the hill on the west. On the top there are features clearly connected with ritual activities. In particular, a building has its windows looking out toward the Aobamba-Santa Teresa ridge and the recently restudied site of Llactapata, where a corresponding construction focuses the view on Huayna Picchu [Ziegler, Thomson and Malville 2003]. The highest point is shaped as a sort of arrow which points due south. In this direction of the cardinal cross, the sight at the horizon is blocked by the Salcantay peak, one of the highest mountain of the region of Cusco (6271 m). Salcantay was certainly revered already in Inca times and is today perceived as the brother of the (slightly higher) Ausangate, the highest peak of the Inca heartland, located east of Cusco.

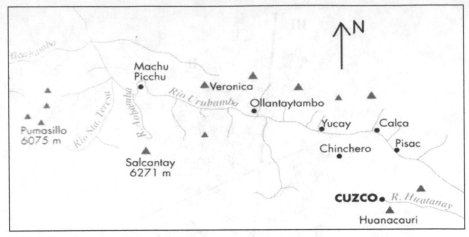

Fig. 4. The geographical position of Machu Picchu with respect to "cardinal" mountains and to the Urubamba river

Important mountains are also located in correspondence to the east and the west of Machu Picchu. Indeed if we take the (slightly elevated) point of view of an observer located on the Intihuatana platform, looking east we can immediately discern one of the two "pyramidal" peaks of Mount Veronica (5750 m). To the west instead, the peaks of the Pumasillo (6075 m) range span the southwestern horizon, with the northernmost end of the range located due west.

Interestingly, in accordance with the fact that everything in Inca religion had dual, complimentary aspects, the relationship of Machu Picchu with the mountains is no exception. In fact, the mountains, with respect to which the town occupies such a "special place", are "replicated" in many sculpted "image rocks" scattered in the town itself.

Finally, the position of Machu Picchu must also certainly be considered "special" with respect to the Urubamba (Vilcanota) River, since – as already mentioned – the river makes a three-sided turn around the Huayna Picchu-Machu Picchu ridge, which appears therefore as a sort of peninsula marking a neat topographical end to the valley.

To give an overview of the relationships of Machu Picchu with the sky – i.e., of the archaeoastronomy of Machu Picchu – it is necessary to distinguish neatly between the "built", residential part of the town, and therefore the eastern flank (sectors I-II) and the "ceremonial" part located on the western flank (sector IV). Indeed, in sectors I-II a clear interest in accurate celestial observations has been documented in many buildings:[2]

1) The so-called Torreon: a P-shaped building of fine Inca masonry which encircles a shaped stone. The motion of the rising sun near the winter solstice could be followed here by means of the shadows cast by a cord affixed at the window [Dearborn and White 1989]. Also the Pleiades and Scorpio were probably observed here.

2) The underlying cave (customarily called the Royal Mausoleum) beneath the Torreon, also aligned to the June solstice sunrise [Malville and Ziegler 2007].

3) The cave usually called Intimachay: a natural cavity with a carefully cut tunnel-

window which allows a quite precise measurement of the sun rising at the summer solstice [Dearborn, Schreiber and White 1987]

4) A general "solstitial" planning of the whole "royal" sector.

People living in the residential sectors had therefore access to devices aimed at following the timing of the solstices with good precision. Vice-versa, only a generic, rather symbolic interest in the sky is recognizable in the ceremonial sector. Actually, and in spite of several existing claims (regarding for example the astronomical function of the Intihuatana stone) to the best of this author's knowledge no clear interest in *precise* astronomical measurements has ever been convincingly documented here. It is therefore interesting that astronomical sightings were aimed at the *rising* of the sun and/or other celestial bodies, while the likely zone where rites were carried out seems to be suited for "popular" observations of phenomena at *setting*. In particular, from the Intihuatana the highest summit of Pumasillo roughly corresponds to the azimuth of the setting sun at the December solstice while the sun at the equinoxes is seen as setting at the northern end of the same range.

Finally, to the possible astronomical observations carried on from Machu Picchu we should add (as noted by Reinhard [2007]) the fact that, since the culmination of the sun between the two zenith passages and that of all the stars of the southern portion of the sky obviously occurs due south, the imposing mountain of Salcantay could be used as a useful, distant foresight.

The place were Machu Picchu was built therefore seems to have been chosen because it satisfied an impressive number of geographical/symbolical requirements. It is, of course, possible – although unlikely – to think that such relationships are due to chance, or that they were ancillary and are not a fundamental key to explain the choice of the site. However, the position of Machu Picchu also exhibits *another* interesting feature. To discuss it, we observe that, together with the cardinal directions, the directions characterized by a relatively thin "void" of azimuths between 135° and 155° degrees (as measured from Cusco) are of tantamount importance in understanding the complex connections between religion, astronomy, cosmology, and sacred geography among the Incas [Urton 1978; Zuidema 1982a; Magli 2005].

First of all, it must of course be observed that this "void" characterizes the "preferred" direction of the orography of the geographical region of interest here, which spans some 500 km. starting from Tiahuanaco, following Lake Titicaca and the course of the rivers up to the Viracocha temple in Ratqui, then Cusco (the Inca heartland) and, finally, bends slightly west to follow the Urubamba valley. Machu Picchu stands, in a sense, at the ideal end of this corridor since the river makes a complete turn around its ridge.

As we shall explain in detail later on, this "void" – clearly not by chance – also characterized the "ideal direction" of the Inca cosmological myth. These two aspects merge with yet another very important one, namely, the fact that the "void" happens to be connected with orientation. Indeed, in the southern hemisphere a "pole star" is *never* available (since precession never brings the south celestial pole sufficiently close to a bright star). As a consequence, the natural way to establish (roughly) the position of the south pole is – and has always been – to follow a bright star or group of stars up to culmination. The most natural choice at the Cusco latitudes is to follow those bright stars of the Milky Way which culminate relatively near the pole, and indeed the principal constellations of the Incas were located along the Milky Way. Among them, particularly relevant for orientation were the stars of the constellations called Crux-Centaurus by today's astronomers, and the dark constellation of the Llama (a dark zone of the Milky Way near Centaurus, which looks like the form of a crouching animal) [Urton 1982]. These celestial objects were raising in Inca times (today precession has shifted these values a bit) just in correspondence to the "void"; for instance Gamma-crux and Alpha-crux were rising in 1430 AD at azimuths (viewed from Cusco with a flat horizon) around 146° and 152° respectively, while the head of the Llama was rising roughly between 141° and 151°. It has even been suggested that the rising azimuth of Alpha-crux may have also influenced the choice of the borderline between the two southern *suyus* (parts) of the empire [Urton 1978]. In *any* case, as we shall see more in depth later on, there is little doubt that *Mayu,* the Milky Way, seen as a huge "double branched" celestial river, played a fundamental role in the Inca "cosmovision".

4 Machu Picchu as a pilgrimage center

It has been already proposed by some scholars that Machu Picchu was a *sacred center,* visually and symbolically connected with several other huacas of the region and was therefore a destination of pilgrimages. In particular, this idea inspired Reinhard [2007], and similar conceits also appear as part of the "multi-functional" interpretation discussed by Ziegler and Malville [2007]. In both these works, however, the "royal estate" theory is in any case maintained. As mentioned, it is instead the aim of the present work to propose that Machu Picchu not only was a sacred center, but that it was conceived, designed and built *specifically* to be a place of pilgrimage, the last part of it actually *taking place inside the town.* In this respect, at least to the best of the author's knowledge, the interpretation proposed here can be considered as new.[3]

To develop this interpretation, it is fundamental to start from a comparison between Machu Picchu and the unique Inca pilgrimage site which is historically well documented and has been the subject of a exhaustive study: the Island of the Sun [Bauer and Stanish 2001].

The Island of the Sun is a rocky islet located near the southern end of Lake Titicaca. For some reason that nobody has ever dared to investigate, an apparently insignificant, natural rock formation present in the northern part of this island was identified as nothing less than the place of origin of the sun, and therefore of the Incas: this place actually appears, although with different details, in most versions of the Inca cosmological myth. As a consequence, a very important Inca sanctuary was located on the island. The whole site was administrated by the state, and the Incas removed the existing population replacing them with colonists from various parts of the empire; also, a specialized group of women was established with the purpose of serving the sanctuary. The sanctuary area of course included the "sacred rock" where the sun was born, which was the final destination of the pilgrims. The pilgrimage took place in successive stages (fig. 5):

1) Pilgrims gathered at what is today Copacabana, and then sailed to the island from the south.

2) Once landed they followed a path oriented – as the island – in a southeast-northwest direction. The path ultimately brought them to the most sacred part, which was enclosed by a (low) wall.

3) Apparently not all of the pilgrims were allowed to pass the entrance to the

northern part of the island; those not admitted could still have a look anyway at the rites from terraces appositely leveled out of the wall.

4) After the wall, the sacred path passed by some other gates and "stations". In particular, near the building today called Mama Ojlia, the pilgrims could look at the "footprints of the sun", an area where the exposed bedrock contains natural marks resembling huge footprints.

5) Finally, people gathered in the plaza in front of the Sacred Rock, where they witnessed rites and, at the time of the winter solstice, the hierophany of the sun setting between two pillars on a ridge to the northwest.

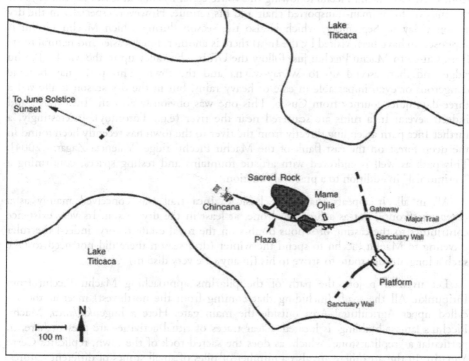

Fig. 5. Map of the Inca sanctuary on the Island of the Sun. The position of the pillars used to observe the winter solstice sunset from the sanctuary area is indicated. After [Bauer and Stanish 2001]

Bauer and Stanish describe the layout of the site as follows:

It is not by chance that the final destination of the pilgrims was on the point of land farthest from the mainland, on the northwest side of the island of the Sun. The sanctuary was, like many pilgrimage centers of the world, situated in a remote location that served to emphasize its otherworldliness [Bauer and Stanish 2001: 247].

Clearly, the very same words may be applied to Machu Picchu, which is located in a similar way: it suffices to substitute the lake surrounding the Island of the Sun with the three-sided gorge of the Urubamba River. Actually, as we shall see, the list of similarities is much longer and more impressive than this. Let us, indeed, turn our attention to Machu Picchu again.

There are two main ways to approach Machu Picchu coming from the Inca heartland. The first one is the world-famous, spectacular route – usually called the *Inca trail* – which departs from the Urubamba River at Pikillacta (Llactapata), some 30 km. from the Machu Picchu ridge. It leads to the Inca site of Winay-wayna and then to the so-called Intipunku, a building clearly meant to provide a monumental, controlled access and located south of Machu Picchu, in plain view of the town.

The "Inca trail" is indeed certainly Inca, since it is in many points carved in the rocks and passes through or near several Inca settlements. However, it is a very uneasy route, rising for example as much as 4200 m to cross the "Dead Woman Pass". The journey from Cusco to Machu Picchu following this route certainly took at least four full days for the Inca on his human-transported chair and his retinue. However, especially in the dry season (May to September, which is also the season during which Machu Picchu is supposed to have been visited by the Inca) there is another, much easier and natural route from Cusco to Machu Picchu: just follow the Urubamba valley up to the Machu Picchu ridge and then ascend up to Winay-wayna and the town. This path may become dangerous or even impassable in case of heavy rains, but in the dry season it allowed a three-day, quiet journey from Cusco. This one was obviously also an "Inca trail", and indeed several Inca ruins are scattered near the river (e.g., Torontoy). Interestingly, a further Inca path ascending directly from the river to the town has recently been found in the deep forest on the east flank of the Machu Picchu ridge [Valencia Zagarra 2004]. This path as well is endowed with artistic fountains and resting spaces, confirming a "cerimonial" in addition to a practical function.

All in all, it appears that the "classical" Inca trail was conceived mainly as a *ceremonial* route, not as a functional one, at least in the dry season. Its very existence contributes to the casting of serious doubts on the royal estate theory: indeed the ruler traveling to Machu Picchu to spend the winter (dry) season there did not need to take such a long, uneasy route to arrive to his (in any case very distant) estate.

Let us now re-join the path of the pilgrims approaching Machu Picchu from Intipunku. All the roads (including that coming from the northwest) meet at the so-called upper Agricultural Area outside the main gate. Here a huge *kallanca*, Machu Picchu's largest building, is located. Clear traces of ritual activities are present here, in particular a "replica stone" which, as does the sacred rock of the town, replicates Cerro Yanatin. In the area there are also a number of piles of small stones of different natures, and at least some were evidently carried to the site from distant places; for instance, a number are rounded river rocks. Probably, thus, at least some of these stones are offerings left upon arrival. The whole area presents clear similarities with the area located outside of the innermost sanctuary on the Island of the Sun as it offers an obstructed view on the western sector of the town and the plaza. We can thus speculate that, exactly as occurred on the Island on the Sun, people who were not admitted further could still look anyway at the events taking place in the town from the platform and the terraces. Curiously, the sector of the Sacred Rock is not visible from here, and perhaps for this reason a "replica" of the same monument was created in this place.

Continuing their way down, the pilgrims eventually reached the gate of the town. The town itself was thus characterized as a *closed space*, and this is unusual for the Incas (actually at Machu Picchu we encounter *most of the few* examples of Inca stone doorways which have devices – rings and recesses – meant for closing them with wooden lintels).

It is in any case clear that the fencing wall was not meant for defensive purposes, since it would have been exceedingly easy for any enemy to block the water channel (which comes in the town from outside) with just a few stones and in Machu Picchu there are no water reservoirs. The aim of the wall is therefore, as we said, to furnish a controlled access and, consequently, to stress the separation of what is inside from what remains outside.

The visitors admitted perhaps left their ritual offerings just near the entrance wall, since many peculiar stone pebbles (mainly of obsidian) have been recovered there. Then, people were confronted by a corridor (service buildings such as stables and magazines are located on the left of the entrance but are separated by a wall) with the imposing view of the Huayna Picchu mountain, the likely final destination of the pilgrimage, just in front of them.

Today, at this point most tourists turn right, visiting the "residential sectors" first. However, in ancient times these sectors – which begin with the Royal residence – were very likely closed to public; therefore, a person entering would have had by necessity to proceed straight in the western sector encountering the sequence of structures we became familiar with in the preceding section: first, the so-called quarry; second, the Temple of the Three Windows and, finally, the Intihuatana platform. But why?

It is my aim here to maintain that sector IV in Machu Picchu, oriented and "scheduled" towards north and Huayna Picchu, was conceived as an image, a *replica* of the path followed by the first Incas in the cosmological myth. This myth has been recounted, although sometimes with different details, in many chronicles. Is starts at the Island of the Sun, from which the first Incas traveled underneath the earth following the "void" (southeast-northwest) direction, emerging at a place called Tampu-tocco. According to Sarmiento de Gamboa [1999] and Cobo [1983], this name means "the house of windows" because "it is certain that in this hill there are three windows" and the first Incas came out from one of these windows. Tampu-tocco is located south of Cusco: the Incas now traveled therefore to the north, up to the summit of the Huanacauri hill, where one of them was turned into stone, becoming a fundamental huaca of the future empire before arriving in the Cusco valley.

There is, at least in my view, little doubt that the key elements of the myth find a close correspondence in the structures of Sector IV. In fact, a newcomer encounters in succession the three main elements of the myth, symbolized respectively by:

1) The quarry. Although it is a true quarrying area, it is also a zone intentionally left in "disorder" located very near the "royal" and the "ceremonial" sectors. It shows signs of ritual activities as well, such as carvings of serpents on the rocks. It is thus tempting to associate it with Pachamama and to think that it was somehow connected with an image of the underground travel of the first Incas.[4]

2) The sacred plaza. Here the Temple of the Three Windows is to be associated with Tampu-Tocco.

3) The Intihuatana pyramid. This might have been conceived as a replica of the Huanacauri hill, the Intihuatana itself resembling in shape the sacred mountain Huayna Picchu located at the end of the path [Reinhard 2007] as well as the sacred stone-huaca which was located on Huanacauri.

It should be noted that, inspired precisely by the Temple of the Three Windows, Hiram Bingham proposed that the town had to be *identified* with Tampu-tocco. It is *not* my intention here to revive his theory. In fact, there is no doubt about the identification of the Huanacauri hill as one of the most important huacas of the Cusco ceque system [Bauer 1998] and actually most chronicles associate Tampu-tocco with the Pacariqtambo hills south of Cusco (the imposing Inca site of Maukallakta was probably built there to recall the mythical events, see [Bauer and Stanish 2001]). The idea here is rather that Machu Picchu was conceived, as many other sanctuaries around the world, as a powerful and tangible replica of the holy places of the myth. It is in fact well known that sanctuaries are often built to offer "replicas" related to the same sacred event and/or place in different sites and at different stages of monumentality (think, for instance, of the various "copies of the Holy Land" constructed in Europe during the Middle Ages and the Renaissance). It is an easy process to render such places sacred as well, for instance with the presence of holy images, relics, or oracles.[5] In this connection, note the following passage appearing in the chronicle of Bernabè' Cobo:

> The Incas had founded a town on the site of Pacariqtambo, and they built on it, in order to make it famous, a magnificent royal palace with a splendid temple. The ruins of this palace and temple remain even today, and in them some stone statues and idols are seen. At the entrance of that famous cave at Pacariqtambo there is a carefully cut stone window in memory of the time when Manco Capac left through it [Cobo 1983].

This passage is somewhat strange. First of all, it hardly fits with what is visible – at least today – in Pacariqtambo, while it could be easily referred to Machu Picchu. Further, it gives the impression that the author is speaking about a place where he has never been. It is actually difficult to believe that "statues and idols" could have been left on a site so near to Cusco and known to the Spaniards at the time Cobo was writing.

The connection between "power and replica" is especially true for those sanctuaries which were aimed at the foundation of the temporal power in Durkheim's [1912] sense, such as, without a doubt, those administered by the Inca state. Again, the best known example is that on the Island of the Sun, where cyclical ritual activities took place at scheduled times during the year, with the culmination at the winter solstice hierophany. It is thus tempting to attribute a similar role and function at Machu Picchu as well, and to conclude that the place was administered by a dedicated group of "priest-astronomers" (*amautas*) in charge of "controlling" the whole course of the *pacha* – the sacred "space and time" of the Incas – as shown by the astronomical alignments of the buildings of the residential sectors (which include *both* solstices).

Finally, although only at the level of what may be simply an interesting coincidence, it can be noted that there are impressive similarities between the NW sector of Machu Picchu and some seventeenth-century drawings appearing in the so-called Miccinelli Documents [Laurencich-Minelli 2006, 2007] and in the Poma de Ayala Chronicle. Actually, as already noted several times in the literature (see e.g. again [Reinhard 2007]) the representation of the Huanacauri idol in the rites of the Inca month of March depicted in this document closely resembles the Intihuatana (fig. 6, left). Of course, many huacas of this type existed before the conquest; however, the resemblance becomes more impressive if we also consider the Poma representations of Tampu-tocco (fig. 6, right). This place is indeed shown with its three windows at the base of the Huanacauri hill. Perhaps the choice of the month is not casual either: March is the only case in which the Inca name of the month mentioned by Poma exhibits the suffix -*pacha* [Laurencich, Minelli and Magli 2010]. The equinox was a moment at which the sun (*hanan*) and the moon (*hurin* – thought of as representing the dark hours) were equilibrated having the same "strength"; in this month sacrifices to Viracocha were made and, according to ethnographical data, the full moon before the equinox was the time of a feast devoted to the hanan dead, who were in turn associated with the summit of the mountains.

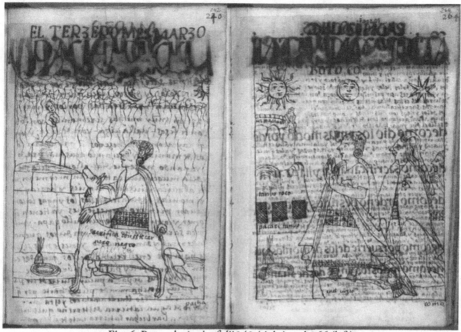

Fig. 6. Poma de Ayala *folii* 242 (right) and 266 (left)

5 Discussion

As explained in detail in Appendix 1, any interpretation of Machu Picchu is destined to remain speculative unless we eventually find out for sure how the town was called by the Incas and discover written texts mentioning both the town and its function in an explicit way. Clearly, however, the interpretation proposed in the present paper would be strengthened if it helps to explain why the town was built in such an accurately chosen, special position. We are thus led to investigate if it is possible to identify a link between topography on the earth and celestial cycles in the sky at the times of the Incas.

As a general observation, it *must* be noticed that these kinds of issues are extremely delicate in archaeoastronomy in general. Actually, very few of the proposed examples of monuments or human-built landscapes anywhere in the world associated with "terrestrial images" of the sky are securely proved (see [Magli 2009] for a complete discussion). However, in the case of the Incas we have a quite solid starting point which relies on the already mentioned fieldwork of Urton [1982] among the contemporary Misminay people. This research has made particularly evident the fundamental role played by the Milky Way. In the centuries before the Incas, precession brought the solstitial points near the intersections between our galaxy and the ecliptic (fig. 7).

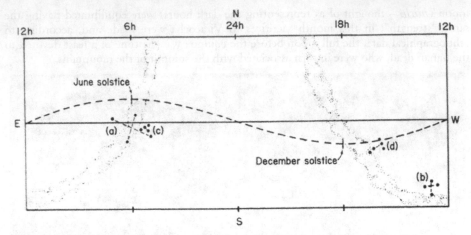

Fig. 7. The intersection of the Milky Way with the Ecliptic. The positions of the solstices in the Milky Way, respectively in Sagittarius and near Gemini, are shown. Four star-to-star "cross" constellations of today's Misminay people are also shown: a) a combination of stars in the area of Gemini/Orion; b) our Southern Cross; c) the Belt of Orion plus probably Rigel; d) five stars in the head of Scorpio. From [Urton 1982]

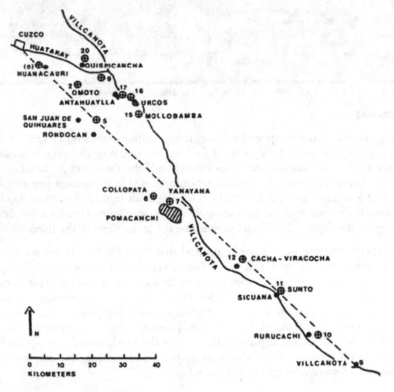

Fig. 8. The pilgrimage to the Vilcanota Hill according to Zuidema [1982]

As a consequence, the position of the sun with respect to the Milky Way could be used to estimate the times of the solstices, which were, in turn, fundamental dates among the yearly Inca rituals.[6] The bright arc of our galaxy in the sky was perceived as a "celestial river" having a "terrestrial counterpart" in the Vilcanota River, which, in Urton's words, is "equated with the Milky Way". This is, therefore, a hint to the possibility that Machu Picchu, located in such a special position with respect to the Vilcanota river, may have had a special significance connected with the sky as well. Actually, written documentations of two Inca pilgrimages along the course of the Vilcanota exists in the work of the chronicler Cristobal De Molina. The first (studied in detail in [Zuidema 1982b]) moved south of east from Cusco and was connected with the cosmological myth (this pilgrimage was in all likelihood a *Capac-Cocha*, namely, it ended with a human sacrifice). It occurred at the *winter* solstice and was directed "toward the place where the sun was born" (fig. 8). The final station was a huaca located on the summit of the Vilcanota hill, some 150 km. as the crow flies from the capital at the border between the Vilcanota and the Titicaca watershed. Coming back, the priests followed the course of the river, bringing offerings to other huacas. The second pilgrimage was instead directed north of west, i.e., in the opposite direction, and occurred a few weeks after the *summer* solstice. During the night, illuminated by torches, runners started from Cusco and reached a bridge in Ollantaytambo, where Coca leaves were offered to the river.

In both cases, a fundamental role was played by three key elements: the opposite solstice timing, the opposite "void" direction, and the Vilcanota river alongside it. The two pilgrimages thus took place on the earth, along the terrestrial image of the Milky Way, towards two "ends" which – due to the timing – were clearly connected with the two "ends" of the sun's path at the horizon, the journey of our star throughout the fixed stars during the year. Of course, the path of the sun is the Ecliptic, not the Milky Way, but the starting and ending points (the winter and the summer solstice) were both located at the crossroads between the two. Interestingly, the southern pilgrimage ended at the sources of the river, but still far from the "true" place of birth of the sun, the Island of the Sun. Perhaps a similar role was played by Machu Picchu with respect to Ollantaytambo in the case of the northern pilgrimage; actually, the "inter-cardinal" position of the town with respect to the landscape also recalls the presence in the sky of cruciform star-to-star Inca constellations, which were concentrated near the two solstitial points (see again [Urton 1982] and fig. 7).

All in all, the hints of different nature and origin presented in this paper *all* seem to point in the same direction. They suggest that Machu Picchu could have been conceived and built as the ideal – *hanan* – counterpart of the Island of the Sun, in accordance with the duality of the sacred which is typical of the Andean world.[7]

Machu Picchu was indeed located at the ideal, opposite crossroad between the terrestrial and the celestial rivers: the other end of the sun's path.

Acknowledgments

The ideas set forth in this paper benefited very much from several discussions with my friends and colleagues Laura Laurencich Minelli, who taught me the concept of "Inca space-time", and Jean Pierre Protzen, whose comments greatly improved a first draft of the present work.

Appendix 1. Review of the evidence for the "royal estate" theory.

As mentioned in the introduction, an Inca "royal estate" was a land property, nominally owned by the king and managed by his family clan, endowed with agricultural lands and residences for the ruler and the *elite*. The very same interpretation has been repeatedly proposed for Machu Picchu, and it is the aim of this appendix to give what I hope is a complete overview of the evidence – or better, of the almost complete lack of evidence – for this idea.

First of all, to gain any textual information about an ancient place we should know how it was called. As a matter of fact however, we cannot be sure about how the town now known as Machu Picchu was called by the Incas. Indeed, we only know that it lies between two paired peaks which were called by inhabitants of the area – at the moment of Bingham's "discovery" – by the names of *Machu Picchu*, the old peak, and *Huayna Picchu*, the young one. These names were reported to Bingham and he decided to name the town after one of the peaks. However, a town with this name (or, for that matter, with the similar name Picho; see below) is *never* mentioned in the Spanish historical chronicles (at least to the best of the author's knowledge). Of course, although often confused and/or captious, the chronicles constitute *the* fundamental written sources of information about Inca life and history. Therefore, *no* attempt to identify the role and the meaning of the town can be based on historical sources, and any attempt is, consequentially and necessarily, admittedly speculative (this holds of course for the proposals set forth in the present paper as well).

That said, we can proceed to analyze the "royal estate" hypothesis. This hypothesis was formulated by Rowe [1990] starting from the fact that a place called Picho (Picchu) *is* mentioned in some sixteenth-century documents. In particular, it appears in the account of a journey to Vilcabamba. In this manuscript the author, Diego Rodriguez de Figueroa, cites Picho – apparently *without* having visited it – as a place located on a certain path between Condormarca and Tambo. The same place is also mentioned in a legal document of 1562, first discovered by Glave and Remy [1983] where the existence of a Picho *cacique* (local chief) farming coca is also cited. On the basis of this evidence, Rowe proposed that Bingham or his informants had confused the true name of the town – which he thinks was originally called Picho – with that of the nearby mountain.

This interpretation appears reasonable, although it is, of course, difficult to be *certain* of its validity for a series of reasons. First of all, the name Picchu (peak) is *exceedingly common*: for instance, one of the sacred hills northwest of Cusco is called precisely Picchu. Further, it is at least possible that the place was located in the same area of Machu Picchu but down in the valley, along the course of the Urubamba river, to which the administration of agricultural activities referred. Finally, it cannot be forgotten that there is no archaeological evidence for the presence of Spaniards in the town at any time, in spite of existing claims; further, there are no visible signs of intentional destruction of Inca "idols" – a fact exceedingly common in all the Cusco huacas.

In any case, even if we admit that Machu Picchu was called Picchu and therefore mentioned in these writings, we are in any case very far from having any interpretation of the town, and the idea of the "royal estate" has to be justified by other means. To this end, attention is called to the fact that a list of agricultural properties attributes the farmed lands of the Urubamba valley between Ollantaytambo and Chaullay to the private possession of their conqueror, the Inca Pachacuti. Picho is *not* explicitly mentioned in this context, but "since the terrains of the valley bottom belonged to Pachacuti, it is quite probable that the places at higher quotas in the same zone were part of the royal estate of the king as well" ([Rowe 1990], my translation). Again, however, even if admit the validity of this second implication – and of course, we are not obliged to do this – the fact that the citadel was part of the royal properties does not show that it was used as a personal estate of the Inca. This too is, therefore, an inference, and indeed Rowe cautiously concludes "we can suppose [*podemos suponer*] that the Inca ruler choose Machu Picchu as a personal estate and as a memorial of war campaigns in the zone of Vitcos" (my translation). In spite of such correct prudence, this paper is commonly considered (perhaps without its having been read) as giving conclusive proofs for the interpretation of Machu Picchu as a royal estate of Pachacuti. Instead, it is clear that this is only a sort of vague possibility based on as many as *three interdependent implications* (Machu Picchu was called Picho -> Picho was a property of Pachacuti -> Pachacuti built it as his personal royal estate), *none of which is securely proved.*

Of course, together with the written sources, it has to be ascertained if archaeological evidence exists to support the theory. As far as the monumental architecture of the town and its relationship with the landscape and the sky are concerned, this ascertainment is fully developed in the main body of the present paper. Here I will instead mention some recent studies which may be of relevance for the interpretation of the town (see [Burger 2004] and references therein). One is the re-analysis of the bones found by the Bingham expedition. As already mentioned, the percentage of female bones was originally over-estimated; however, a 3:2 excess of female bones remains. It is known that the Incas established a group of women in the sanctuary of the Island of the Sun whose specific role was to serve the sanctuary; perhaps the same might have happened in Machu Picchu. The re-analysis of human remains has also shown that the inhabitants (which should have numbered around 750) exhibited an high degree of population diversity. They therefore came from the different parts of the empire; likewise, a significant part of the pottery recovered in tombs comes from distant provinces. This is relevant for our discussion because it is known that royal estates were staffed by *mitima,* retainers of different ethnic groups removed from their original villages. However, although valid, this observation cannot be used as a proof for the "royal estate" theory, since the very same thing holds for many other Inca state projects and in particular, again, for the sanctuary on the Island of the Sun. Finally, another relevant recent analysis is the reconstruction of the ancient climate at Machu Picchu, which turned out to have been quite similar to that of today: warmer than in Cusco, but also more rainy. It has to be definitively excluded, therefore, that Machu Picchu could have been considered a more pleasant place to stay than the capital during the rainy season (October to April), while a relative improvement of the climate temperature might actually have been obtained by moving to Machu Picchu in winter.

Notes

1. It is worth noting that trials have been made in the past to divide the plan of the town into the two traditional moieties – the upper, or *hanan,* and the lower, or *hurin* – which are typical of the organization of the Andean space. For instance, the capital Cusco was ideally divided according to this principle. Contrary to Cusco, however, where this division is well documented in the chronicles and essentially corresponds to the northern and the southern parts respectively, at Machu Picchu the situation is unclear. In any case, the standard proposal is to identify the *hanan* part as that including the "royal residence" and the quarry (sectors I and IV in fig. 3).
2. The level of precision of such measurements has been the subject of much debate starting from a paper by Dearborn and Schreiber [1986]. This topic is not of specific relevance here however; the interested reader can consult the anthology recently edited by A. Aveni [2008].
3. The possible existence of other functions for the town (for instance, administrative) are not

excluded *a priori*, but have to be considered as subsidiary to the main one. Similarly, although the "royal estate" theory is refuted here, this of course does not mean that the Inca did not visit the town. On the contrary, it is very likely that he did, and the evidence for a "royal residence" is actually quite compelling.

4. A similar "symbolic + functional" interpretation has been proposed by who writes for the famous quarry of Ranu Raraku on Easter Island, where several huge statues were left at different stages of extraction and completion [Magli 2009].

5. It can be noted that important oracular shrines have been documented in the Andean world since extremely ancient times, e.g., in Chavin and Pachacamac, and that some of the caves at Machu Picchu (especially the Temple of the Condor) may suggest the same interpretation.

6. It is worth recalling that the coincidence between the solstices and the intersections of the Ecliptic with the Milky Way was fundamental in Maya cosmology, see e.g., [Schele, Freidel and Parker 1995]

7. People living in the northern hemisphere associate "north" naturally with "up", because their celestial pole – and therefore the center of the apparent rotation of the stars, the place where the *axis mundi* meets the sky – is the northern one. Interestingly, it is apparent that north was associated with "upper" (*hanan*) by the Incas as well, although they were living in the southern hemisphere. The anthropological reasons for this fully merit, at least in the author's view, further investigations.

References

AVENI, A. 2008. *Foundations of New World Cultural Astronomy: A Reader With Commentary.* Boulder: University Press of Colorado.

BAUER, B. 1998. *The Sacred Landscape of the Inca: The Cusco Ceque System.* Austin: University of Texas Press,.

———. 2004. *Ancient Cuzco: Heartland of the Inca.* Austin: University of Texas Press.

BAUER, B., and DEARBORN, D. 1995. *Astronomy and Empire in the Ancient Andes.* Austin: University of Texas Press.

BAUER, B., and STANISH, C. 2001. *Ritual and Pilgrimage in the Ancient Andes: The Islands of the Sun and the Moon.* Austin: University of Texas Press.

BINGHAM, H. 1952. *The lost City of the Incas: The Story of Machu Picchu and its Builders.* London: Phoenix House.

BURGER R. and SALAZAR-BURGER L.C. (Eds) 2004) *Machu Picchu: Unveiling the Mystery of the Incas.* New Haven and London: Yale University Press.

COBO, B. 1983. *History of the Inca Empire* (1653). Roland Hamilton, trans. Austin: University of Texas Press.

DEARBORN, D. S. and SCHREIBER, K. 1986. Here Comes the Sun: The Cuzco–Machu Picchu Connection. *Archaeoastronomy* 9:15–37.

DEARBORN, D.S., SCHREIBER S. and WHITE, R. 1987. Intimachay: A December Solstice Observatory at Machu Picchu, Peru. *American Antiquity* 52, 2: 346-352.

DEARBORN, D.S. and WHITE, R. 1989. Inca observatories: their relation to the calendar and ritual. Pp. 462-469 in *World Archaeoastronomy*, A.F. Aveni, ed. Cambridge: Cambridge University Press.

DE GAMBOA, S. 1999. *History of the Incas* (1572). Clements Markham, trans. New York: Dover Publications.

DURKHEIM, E. 1912. *The Elementary Forms of Religious Life* . Rpt. 1965, London: The Free Press.

GASPARINI, G. and MARGOLIES, L. 1980. *Inca Architecture.* Bloomington: Indiana University Press.

GLAVE, L. and REMY M. 1983. *Estructura agraria y vida rural en una región andina. Ollantaytambo entre los siglos XVI y XIX.* Archivos de Historia Andina. Cusco: Centro de Estudios Rurales Andinos Bartolomé de las Casas.

HYSLOP, J. 1990. *Inka settlement planning.* Austin: University of Texas Press.

KENDALL, A. 1985. *Aspects of Inca architecture: description, function and chronology.* Oxford:

British Archaeological Reports.

LAURENCICH MINELLI L. 2006. UEl mito utopico de Paititi desde un documento jesuitico parcalmente inedito del siglo XVII. *Archivio per l' Antropologia e la Etnologia* CXXXV: 183-202.

———. 2007. *Exsul Immeritus Blas Valera Populo Suo e Historia et Rudimenta Linguae Piruanorum. Indios, gesuiti e spagnoli in due documenti segreti sul Perù del XVII secolo.* CLUEB Bologna.

LAURENCICH MINELLI, L. and MAGLI, G. 2010. A calendar Quipu of the early 17th century and its relationship with the Inca astronomy. *Archaeoastronomy* **22** (in press). Pre-print arxiv.org/abs/0801.1577.

MAGLI, G. 2005. *Mathematics, Astronomy and Sacred Landscape in the Inka Heartland. Nexus Network Journal* **7**, 2: 22-32.

———. 2009. *Mysteries and discoveries of Archaeoastronomy.* New York: Springer-Verlag.

MALVILLE, J. M., and ZIEGLER, G. 2007. Machu Picchu, Inca Pachacuti's Sacred City: A multiple ritual, ceremonial and administrative center. http://www.adventurespecialists.org/mapi1.html. Accessed 23 March 2010.

MALVILLE, J. M., THOMSON, H. and ZIEGLER, G. 2004. El observatorio de Machu Picchu: Redescubrimiento de Llactapata y su templo solar. *Revista Andina* **39**: 9- 40.

NILES, S. 2004. The nature of Inca royal estates. Pp. 49-68 in *Machu Picchu: Unveiling the Mystery of the Incas,* R.Burger and L.C. Salazar-Burger, eds. New Haven and London: Yale University Press.

NILES, S. 1999. *The shape of Inca History.* Iowa City: University of Iowa Press.

PROTZEN, J.P. 1993. *Inca Architecture and Construction at Ollantaytambo.* Oxford: Oxford University Press

———. 2004. The fortress of Saqsa Waman: Was it ever finished? *Ñawpa Pacha* **25–27**: 155-175.

REINHARD, J. 2007. *Machu Picchu: Exploring an Ancient Sacred Center.* New York: Cotsen Institute of Archaeology.

ROWE, J. 1990. Machu Picchu a la luz de documentos de siglo XVI. *Historia* **16**, 1: 139-154.

SCHELE, L., FREIDEL, D., and Parker, J. 1995. *Maya Cosmos.* New York: Quill.

URTON, G. 1978. Orientation in Quechua and Incaic Astronomy. *Ethnology* **17**, 2: 157-167.

———. 1982. *At the Crossroads of the Earth and the Sky: An Andean Cosmology.* Austin: University of Texas Press.

WRIGHT, R. M. and VALENCIA ZEGARRA, A. 2001. *The Machu Picchu Guidebook.* Boulder: Johnson Books.

ZUIDEMA R .T 1982a. Catachillay: The Role of the Pleiades & Southern Cross and Alpha and Beta Centauri in the Calendar of the Incas. Pp. 203-230 in *Ethnoastronomy and Archaeoastronomy in the American Tropics,* Anthony F. Aveni & G. Urton, eds. New York: New York Academy of Sciences, vol. 385.

———. 1982b. Bureaucracy and systematic knowledge in Andean civilisation. Pp. 419-458 in *The Inka and Aztec States 1400-1800,* G.A. Collier, R.I. Rosaldo and J.D. Wirth, eds. Studies in Anthropology. New York: Academic Press.

About the author

Giulio Magli is a full professor in the Faculty of Civil Architecture of the Politecnico of Milan, where he teaches the only official course of Archaeoastronomy ever established in Italy. He earned a Ph.D. in mathematics at the University of Milan in 1992 and his research activity developed in the field of General Relativity Theory, with special attention to problems of relevance in astrophysics. In recent years, however, his research interests have focussed mainly on archaeoastronomy, with special emphasis on the relationship between architecture, landscape and the astronomical lore of ancient cultures. On this subject he has written several papers and the book *Mysteries and Discoveries of Archaeoastronomy,* published in 2005 (in Italian) by Newton & Compton editors, and in English edition by Springer-Verlag (2009).

Tessa Morrison

The School of Architecture
and Built Environment,
The University of Newcastle
Callaghan, NSW, 2308
AUSTRALIA
Tessa.Morrison@newcastle.
edu.au

Keywords: Isaac Newton,
Temple of Solomon,
Vitruvius, Vitruvian man,
Newtonian man
proportions

Research

The Body, the Temple and the Newtonian Man Conundrum

Abstract. From his early days at the University of Cambridge until his death, Isaac Newton had a long running interest in the Temple of Solomon, a topic which appeared in his works on prophecy, chronology and metrology. At the same time that Newton was working on the *Principia*, he reconstructed the Temple and commented on the reconstructions of others. An important part of his investigations concerned the measurements of the Temple, which were harmonic and were built "exactly as the proportion of architecture demands." Newton considered these proportions to be in accordance with Book III and IV of *De Architectura*. However, while insisting on exact architectural proportions, Newton moved away from the traditional proportions of the Vitruvian man; he derived a Newtonian man. This poses an interesting conundrum: Newton accepted the Temple's architectural proportions as outlined in Vitruvius's Book III, yet he rejected the human model Vitruvius used as the foundation of these proportions. At the same time Newton accepted the human frame as the basis of all ancient measurements and attempted to estimate the length of the sacred cubit based on the lengths of the parts for the body and the measurements set out by the ancient writers such a Vitruvius.

The Temple of Solomon, Babson Ms 0434 and 'A Dissertation upon the Sacred Cubit of the Jews'

When Newton died in 1727 he left hundreds of unpublished manuscripts, some dating back to his early days in Cambridge in the 1660s. Newton's heirs invited Thomas Pellett to examine the manuscripts and report on their suitability for publication. After just three days of examining these hundreds of manuscripts, Pellett dismissed the majority of manuscripts as "not fit to be printed" [Gjertsen 1986: 426], "of no scientific value" and as "loose and foul papers" [Manuel 1974: 14]. He only found two sets of manuscripts suitable for publication. The first were two manuscripts on prophecies, and although Pellet claimed that the text on prophecy was imperfect, they were nevertheless worthy of publication. This was eventually prepared for press by Newton's nephew Benjamin Smith [Gjertsen 1986: 399] and published in 1733 as *The Observations upon the Prophecies of Daniel and the Apocalypse of St John*. *Observations* proved to be one of Newton's best sellers in the eighteenth century and it was also translated into Latin and German shortly after its first edition [Hall 1992: 372] .

The second was a set of manuscripts on chronology, which were compiled and arranged by John Conduitt, husband of Newton's niece Catherine, and published in 1728 as *Chronology of Ancient Kingdoms Amended.* The book cannot be considered a success and is exceptionally dull. It is arranged in six chapters, five of which were chronologies of the ancient empires of Greece; Egypt; Assyria; Babylon, and Persia. The other chapter is a description of Solomon's Temple, which is not only an intriguing

Nexus Network Journal 12 (2010) 343–352   NEXUS NETWORK JOURNAL – VOL.12, No. 2, 2010 **343**
DOI 10.1007/s00004-010-0029-1; *published online* 6 May 2010
© 2010 Kim Williams Books, Turin

addition to a book on chronology of ancient kingdoms, but is curiously placed after the chapter on the Babylonian Empire, which destroyed the Temple.[1]

The beginning of the chapter on the Temple is quite dismissive: "The Temple of Solomon being destroyed by the Babylonians, it may not be amiss here to give a description of that edifice" [Newton 1988: 332]. The chapter consists of a brief (barely 3,000-word long) description of its floor plan, with three illustrated floor plans. There is no mention of the style of architecture, its splendour or its significance. The description lacks any enthusiasm and is highly clinical. Its architectural description is problematic and there are parts that do not make structural sense. For example in the *Chronology* Newton claimed that:

> The porch of the Temple was 120 cubits high, and its length from south
> to north equaled the breadth of the House: the House was three stories
> high, which made the height of the Holy Place three times thirty cubits,
> and that of the Most Holy three times twenty [Newton 1988: 342-343].

Since the porch, the Holy Place and the Most Holy of Holies adjoined each other, this description created a strange and confused stepped structure which appears to have no precedents, Biblical or otherwise. The three illustrated floor plans are very detailed, but that detail is not backed up by the text. Furthermore, both the text and the illustrations include an external wall that surrounds the precinct wall, with four gates on the western side, the Gate of Shallecheth, the Gate of Parbar, and the two Gates of Assupim. But these were part of the Second Temple and not Solomon's Temple [2 Samuel 6:11-12].

From this chapter it would be easy to conclude that Newton had no knowledge of architecture and that his interest in Solomon's Temple was only as a biblical symbol and, with its destruction, an important historic event. However, the converse is true. Not only did Newton have a good working knowledge of Vitruvian theory, he had a long running interest in the Temple of Solomon that spanned over fifty years.

Over this fifty years Newton wrote many manuscripts that related to the Temple (for example, [Newton undated(a), undated(c), ca. mid-1680s and ca. 1690s]). Two manuscripts are of primary interest for this paper. The first was written in the mid-1680s and was entitled by Newton "Introduction to the Lexicon of the Prophets, Part two: About the appearance of the Jewish Temple", more commonly known by its call number Babson Ms 0434. The other is an earlier unpublished manuscript of the 1680s, an appendix entitled "De magnitudine cubiti sacri" [Newton c1680s(b)], which is a part of a draft on Solomon's Temple that developed into Babson MS 0434. In 1737 this appendix was translated from Latin into English and published as "A Dissertation upon the Sacred Cubit of the Jews", in *Miscellaneous Works of John Greaves Professor of Geometry at Oxford* [Newton 1737].

"Dissertation" is a work of metrology. Newton examined the measurements taken by John Greaves (1602-1652), Savilian Professor of Astronomy at the University of Oxford, who conducted a survey of the Pyramids of Giza which resulted in the publication of *Pyramidographia* in 1646. Greaves's measurements in English feet, taken at the Pyramids, were used to calculate the Royal cubit, Memphis cubit and the Egyptian cubit. From Greaves's calculations of the ancient cubits, Newton proceeded to calculate the measurement of the Jewish sacred cubit, which was essential to understanding the Temple structure.

Newton's "Dissertation" begins: "To the description of the Temple belongs the knowledge of the sacred cubit; to the understanding of which, the knowledge of the cubits of the different nations will be conducive" [Newton 1737: 405]. Newton used Greaves's measurements of the Great Pyramid and systematically compared them with measurements given by ancient sources such as Herodotus, Vitruvius, Strabo, Josephus, Hesychius of Alexandria, Lucius Iunius Moderatus Columella, Philandrier, Gnaeus Julius Agricola, Publius Clodius Thrasea Paetus, the Talmud and more contemporary writers such as Willebrord Snellius, Samuel Purchas and Juan Bautista Villalpando. Newton also cited Arabic sources, such as Ibn Abd Alhokm (321-405) [Newton 1737: 408].

In Babson Ms 0434 Newton systematically reconstructed the Temple of Solomon. His primary source for his reconstruction was the Book of Ezekiel. However, Newton examined the changes in the Temple over time. The building of the second Temple by Zerubbabel followed the same foundations but with a great deal less grandeur. It had the same dimensions and was a pragmatic house of worship, but its architecture was mundane and it was nothing to look at [Newton c1680s(a) 5r]. Cyrus the Great ordered the building of the Temple and the internal atrium but nothing else was added. This was the sanctuary that was maintained up to the time of Alexander the Great, as reported by the pagan writer Hecataeus. The Temple was further fortified under Simeon the Just, until Herod built a more sumptuous building for the sanctuary. According to Newton, "God, predicting all these things, thus he corrected them through the prophet Ezekiel" [Newton 1680s(a): 7r]. But Ezekiel left out details of the building and the Angel who revealed the Temple and its measurements to Ezekiel did not show him the entire Temple. Thus by examining the architectural features through time and with the writing of ancient writers, such as Philo, Hecataeus, Josephus, Maimonides, the Talmud and the Septuaginta, Newton was able to reconstruct the Temple of Solomon, removing the features that had been added by the later builders. He stated "we complete the description of the Temple [of Solomon by] comparing all the Temples to each other and supplying what Ezekiel omitted relative to the Temples of Solomon and of Herod" [Newton c1680s(a): 59r]. From this description of the Temple, Newton claimed that it is possible to distinguish the plan of the Temple of Solomon. Since Zerubbabel had built on the foundations of the Temple of Solomon, everything that Zerubbabel and Herod added, or anything that is irregular, must be rejected. Symmetry and harmony in the design of the Temple were important factors in the layout of the Temple plan. He stated that, "The structure is made valuable by such great simplicity and harmony of all its proportions" [Newton c1680a: 65r].

Babson Ms 0434 is a working document; it is incomplete and it contains two reconstructions, the second of which, the more detailed of the two descriptions, is a refinement of the first. The illustration of the Temple precinct in Babson Ms 0434 is in fact the first reconstruction and the second reconstruction is only verbally expressed but in sufficient detail to reconstruct it. In both reconstructions, symmetry was of paramount importance to the floor plan.

Newton's knowledge of Vitruvius, the measurements of the body and the Jewish cubit

In his reconstruction Newton not only outlined the structure of the Temple, he examined the colonnades: the numbers of columns, their height, their thickness, their intervals and their style. These he claimed were determined according to the proportions of architecture. Newton revealed that he was familiar with the architectural theory of

Vitruvius's *De Architectura*, particularly Book III and IV. When Newton derived the width of the intercolumniations from the measurement of the column given by Josephus, he paraphrased *De Architectura* Book III, Chapter III, "The Proportions of Intercolumniations and of Columns". He stated, "The intervals of these pedestals, according to the proportions of architecture, should not be less than the pedestals" [Newton c1608s(a): 29r]. From Vitruvius, Book IV, Chapter III and the measurements of Josephus, Newton estimated the height of the columns as being "six times the thickness according to the Doric style" [Newton c1680s(a): 36r]. In Ezekiel 40:14 the measurement of the height of the doorway is given as twenty cubits; thus Newton concluded that "the width of the doorway was of ten cubits and the height according to the rules of the architects, should be double the width" [Newton c1680s(a): 45r; Vitruvius 1960: IV, vi, 6]. For Newton most of the measurements of the Temple are "exactly as the proportion of architecture demands" [Newton c1680s(a): 10r]. However, according to Newton the architecture of the Temple sometimes surpassed the beauty that classical architecture demands. He confirmed that there was a row of twenty-one columns and twenty inter-columns in the Royal colonnade from the measurement described by Josephus. Newton stated:

> the Royal colonnade will occupy seventeen, twenty or twenty-four [spaces] between the columns of the same magnitude. But seventeen, according to the architectural proportions, will be too few, and twenty-four will be excessive if the columns were estimated to be equal to those of the other atriums, and, in one and another case, are set too far apart by the numbers of Josephus, therefore it should be twenty [intercolumniations]. According to this proportion, the columns will be less numerous than in the proportion of the eustyli of Vitruvius, but more beautiful; and here, where instead of the architrave there are large blocks of marble that cannot be broken, it does not fit the objections of Vitruvius [Newton c1680s(a): 37r] (this and all other quotes from Babson 0434 are translated from the Latin by the author).

Harmony and symmetry in the design of the Temple were important elements in the layout of the Temple plan. Any element that was described by the ancient writers that was irregular had to be rejected. He stated that, "The structure is made valuable by such great simplicity and harmony of all its proportions" [Newton c1680s(a): 62r]. The perfection of the measurements was of paramount importance to the design.

In Babson Ms 0434 Newton evaluated the measurements, revealing their proportional perfection. In the Book of Ezekiel, the prophet is guided by an angel measuring each part of the Temple as they move around the Temple precinct. The measurements of the Temple were in Jewish cubits. There were two types of Jewish cubits, described by Newton as the sacred cubit and the vulgar cubit. The description of the cubits in Ezekiel is very confusing as he claimed that, "The cubit is a cubit and a palm breadth", leaving the distinction between the two cubits ambiguous. In 2 Chronicles 3:2, Solomon instructed that the Temple be built in cubits "after the first measure". A cubit was equal to the length of the forearm from the elbow joint to the end of the middle finger. This simple measurement is inscribed in Egyptian hieroglyphs. The hieroglyph for a cubit is the image of the forearm [Glazebrook 1931: 413].

In "A Dissertation upon the Sacred Cubit of the Jews", Newton not only attempted to resolve this ambiguity, he also attempted to estimate the length of both cubits. He quoted Vitruvius's measurement of the Roman and Greek cubits as being one and a half Roman feet [Newton 1737: 405; Vitruvius, 1960: III, i]. A Roman foot is 0.97 of an English foot. Newton examined both the Roman and Greek cubits and feet; measuring them in palms and digits, together with the Greek *orgyiae*, the span of the arms fully outstretched, because these measurements were defined by the ancient authors. To estimate their value Newton approached the problem of the variations in the measurements of the ancient authors by assessing the limits of each and then comparing them to each other.

Newton reasoned that the builders of the Great Pyramid would have used a uniform unit of measurement in their design, which would have been the ancient measurement of a cubit. In his calculations he claimed that one Greek orgyiae is equal to four Memphis cubits. After converting the measurement of Greaves from English feet to Roman palms and digits he then compared them with the measurements of the ancients. From this Newton concluded, "And it is my opinion that the Pyramid was built throughout after the measure of this (Memphis) cubit" [Newton 1737: 413]. Newton supported this argument that the ancient buildings were built to a standard unit of measurement by considering the measurement of Babylonians bricks. They were all uniform in size, according to the measurements of sixteenth-century travel writer Samuel Purchas, their length was one foot, the width was eight inches and the thickness was six inches. Thus the combination of two brick lengths, three brick widths and four brick thicknesses formed square cubits.

Newton claimed that all measurements which exceeded human proportions, such as the Roman *calamus, clima, scruplum, actus* and many others, were deduced from the multiples of human proportions. The ancient nations rounded off their large numbers into even numbers of cubits – the cubit of man [Newton 1737: 416]. Newton derived the length of many nations' cubits: Memphis, Egyptian, Greek, Roman, Arabian and Babylonian. Although the different lengths of these cubits conformed to the cubits of man, with the exception of the Babylonian cubit which he estimated at two English feet [Newton 1737: 414], this may be an error of inscription, since later in the paper he referred to the Babylonian cubit as being two Roman feet, but both measurements are larger that the human elbow; still, he continued to maintain that the Babylonians built in cubits.

Greaves found that the modern Egyptian cubit was 1.824 English feet, exceeding that of the ancient Egyptian cubit or Memphis cubit. "The measurements of feet and cubits now exceed the proportion of the human members" [Newton 1737: 417]. According to Greaves's measurements of the Egyptian monuments, the human stature was the same as it was in ancient times. The measurements had increased in length because of human and instrument error.

> Feet and cubits were first used (as a measurement) in every nation according to the proportion of the members of a man, from which they were taken. For the foot of a man is to the cubit or lower part of the arm of the same man as about 5 to 9 [Newton 1737: 419].

Newton confirmed this ratio 5:9 between the foot and cubit with other ancient measurements. He considered that the Jewish measurements were determined in the same manner.

He claimed that the Jewish vulgar cubit cannot exceed the cubit of a tall man. Newton claimed that:

> The stature of the human body, according to the Talmud, contains about three cubits from the feet to the head; and if the feet be raised, and the arms lifted up, it will add one cubit more and contain four cubits. Now the ordinary stature of men, when they are barefoot, is greater than five Roman feet, and less that six Roman feet, and may be best fixed at five feet and an half [Newton 1737: 421].[2]

According to Erubin 48a in the Talmud, the area of "his place" is "three cubits for his body and one cubit to enable him to take up an object at his feet and put it down at his head." Newton moved away from the classical "Vitruvian" man. In Vitruvius, the height of man is set at six Roman feet; Vitruvius claimed that the number six was perfect and this perfection was further expressed in the Roman cubit, which equaled six palms or 24 digits, but to Newton and the Bible six palms was a sacred Jewish cubit. Newton's measurements of the stature of a man, five to six Roman feet, equaled three vulgar Jewish cubits of five palms each cubit; thus a vulgar cubit was to be no less than 20 Roman unciæ[3] and no more than 24 unciæ,[4] also from this measurement the sacred Jewish cubit, of six cubits, he calculated to be no less than 24 Roman unciæ and no more than 28.8 unciæ.

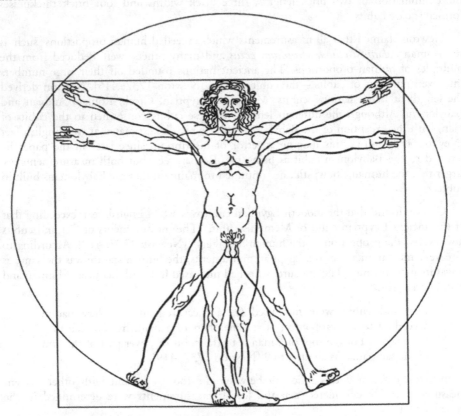

Fig. 1. The Newtonian Man

Newton gave two examples from ancient literature, where he further defined the limits of the sacred cubit. In the first, Josephus wrote that the columns of the great court of the Jewish Temple could be embraced by three men with their arms joined. Vitruvius stated, "For if we measure the distance from the soles of the feet to the top of the head, and then apply that measure to the outstretched arms, the breadth will be found to be the same as the height" [Vitruvius 1960: III.i.3]. However, Newton claimed that although the *orgyia*, or the length of the outstretched arms of a man, was supposed to be the same as the height of a man, in fact it was a palm wider [Newton 1737: 422]. Newton abandoned the traditional image of the Vitruvian man, which is circumscribed by the circle and the square, by adding an extra palm to the length of a man's outstretched arms, giving a slightly more elliptical and rectangular image to the geometry of man (fig. 1). The circumference of the columns, according to the Talmud and Josephus is eight cubits; for Newton, this is equal to three times the height of a man plus three palms, i.e., greater than 15.75 Roman feet and less than 18.75 Roman feet. This further defined the sacred cubit as greater than two Roman feet and less than two and a third Roman feet.

In Newton's second example of the use of the cubit from the ancient literature, the Sabbath-day's journey, in the opinion of what Newton called the 'unanimous' content of the Talmud and all the Jews, was two thousand cubits. According to Josephus, this measurement is not so consistent; in one place he claimed that the Sabbath-day's journey is five stades (three thousand Roman feet) and in another place, six stades (three thousand-six hundred Rome feet) [Josephus 1963: V.2.3; XX.8.6]. Newton, who was very familiar with the work of Josephus, used the reference from the Talmud instead and claimed that instead of "cubits" the Jews sometimes substituted "paces". Walking on the Sabbath is not hurried but is of a moderated speed: "Now man of a middling stature, in walking in this manner, go every step more than two Roman feet, and less that two and a third. And within these limits was the sacred cubit circumscribed" [Newton 1737: 424].

Turning to Vitruvius for the correct architectural height of a step [Vitruvius 1960: III.iv.4], Newton claimed that the middling proportion referred to by the Jews was about 13.5 unciæ and from this he calculated that a pace or sacred cubit was more that 24 unciæ and less than 27 unciæ. From the examples of the height of a man, the circumference of the columns and the Sabbath-day's walk, Newton defined the limits of the sacred cubit and rejected "the erroneous opinions of other writers". Newton concluded that the vulgar cubit was five palms, the cubit of man was equal to 21.4 unciæ or 1.717 English feet, while the sacred cubit was six palms [Newton 1737: 427], the cubit of man plus a palm, was equal to 25.6 unciæ or 2.068 English feet.

The conundrum of the Newtonian man

Unlike many commentators of his time Newton does not directly include or refer to any anthropomorphic element in his reconstruction of the Temple, where the figure of man/God was reflected in the measurements and geometry of the Temple, which prefigured the perfection of the mystical body of the Church. While Newton insisted on exact architectural proportions, he moved away from the traditional proportions of the Vitruvian man, which had been an important element in other contemporary reconstructions. This poses an interesting conundrum: Newton accepted the Temple's architectural proportions as outlined in Vitruvius' Book III yet he rejected the human model Vitruvius used as the foundation of these proportions. At the same time, Newton accepted that the human frame was the basis of all ancient measurements, and he attempted to estimate the length of the sacred cubit with the lengths of the parts of the body and the measurements of ancient writers, such as Vitruvius.

When writing Babson Ms 0434 and "Dissertation upon the Sacred Cubit of the Jews" he was at the height of his intellectual power and was completing the first edition of the *Principia*. The rejection of the Vitruvian man as a model for the proportion of the Temple cannot be dismissed as an oversight by Newton. Newton was aware of Book III of *De Architectura* and the image of the Vitruvian man was also well represented in architectural text. Furthermore, the images of the cosmological/Vitruvian man were strongly linked in the Renaissance with the rise of Hermetic philosophy; and in Newton's unpublished papers he demonstrated an interest in the symbolism of Hermetic philosophy [Newton undated(a), undated(b), undated(c); Anonymous (Trismegist) 2002]. The framing of the model from the Book of Erubin in the Talmud and adding a palm to the span of a man's outstretched arms must have been a conscious alternative.

Conclusion

Newton's manuscripts on the Temple span over fifty years, and the majority of these papers are theological rather than architectural in nature. However, the architecture of the Temple plays an important role in Newton's work on the language of the prophets. The prophets could only be interpreted through "hieroglyphs" [Newton 1957] and one of those hieroglyphs was the framework of the architecture and rituals of Solomon's Temple. This view of the Temple is not only confirmed by his unpublished papers but also by Newton's title for Babson Ms 0434, "Introduction to the Lexicon of the Prophets, Part two: About the appearance of the Jewish Temple". Newton believed that the ancient religion, which he claimed was the original religion of God, understood the mathematical principle of God's orderly design which sustained the solar system. He perceived that they had a pure knowledge of the workings of natural philosophy [Newton c1690s]. The symbol of the Temple was important to Newton, and he returned to the topic many times over the fifty year period. At first he refined it, but eventually, towards the end of his life, he sanitized his work. Before his death, Newton was preparing the *Chronology of Ancient Kingdoms Amended* for publication; by then he had a legacy to maintain, and to maintain it he sanitized a lot of his work, disguising his religious beliefs [Westfall 1980: 817] which at the time were heretical and would have seen him publicly disgraced.[5] The chapter on the description of Solomon's Temple in the *Chronology* had become so sanitized that it is virtually nonsensical and had lost the brilliance that characterised his early work.

The conundrum of the Newtonian man is an interesting puzzle but it is one that has no solution, for Newton left no clue as to why he moved away from the traditional image of the Vitruvian man. However, what Babson Ms 0434 and "A Dissertation upon the Sacred Cubit of the Jews" does reveal is that he had a good working knowledge of Vitruvian theory and an interest in architectural aesthetics, are two aspects of his character that are not normally associated with one of the greatest scientists in history.

Notes

1. There has been some excellent research carried out on the theological implications of the role of the Temple of Solomon in Newton's work (see [Mandelbrote 1993 & 2007; Faur 2004; Goldish 1998]). However, none of these have examined Newton's knowledge of architectural and Vitruvian theory. In particular, "A Dissertation upon the Sacred Cubit of the Jews" discussed in this paper, although frequently referenced as proof of Newton's interest in the

length of the Jewish cubit (for example [Leshem 2003: Popkin 1992; Westfall, 1980]), is a neglected paper that has not been discussed in its own right.
2. Newton's references to the Talmud are incorrect. His reference is Mishnaioth, Tract. De Ghaburim, cap. 4; it should be Talmud, Erubin, 48a.
3. An *unciæ* is a Roman inch, 20 unciæ equals 1.612 English feet.
4. 24 unciæ equals 1.934 English feet.
5. William Whiston, former pupil and successor to Newton as Lucasian Professor at Cambridge, also shared Newton's beliefs. However, he made his beliefs public, which ended his career at Cambridge. He was later charged with heresy and, although not convicted, he never held an academic position again [Force 1985].

References

ANONYMOUS (Hermes Trismegist). 2002. Tabula Saragdina. Pg. 274 in *The Janus Faces of Genius*, B. J. T. Dobbs, ed., Isaac Newton, trans. Cambridge: Cambridge University Press, 274.

BIRCH, Thomas. 1737. *Miscellaneous Works of John Greaves*. London: Printed by J.Hughs for J. Brindley.

FORCE, James E. 1985. *William Whiston*. Cambridge: Cambridge University Press.

FAUR, Jose. 2004. Newton, Maimonidean. *Review of Rabbinic Judaism* 6, 2/3: 215-49.

GLAZEBROOK, Richard T. 1931. Standards of Measurement, Their History and Development. *The Proceedings of the Physical Society* 43.

GJERTSEN, Derek. 1986. *The Newton Handbook*. London and New York: Routledge & Kegan Paul.

GOLDISH, Matt. 1998. *Judaism in the Theology of Sir Isaac Newton*. London: Kluwer Academic Publishers.

GREAVES, John. 1646. *Pyramidographia: Or, a Description of the Pyramids in Aegypt*. London.

HALL, A. Rupert. 1992. *Isaac Newton: Adventurer in Thought*. Cambridge: Cambridge University Press.

JOSEPHUS. 1963. *Jewish Antiquities*. 9 vols. London: William Heinemann Ltd.

LESHEM, Ayval. 2003. *Newton on Mathematics and Spiritual Purity*. International Archives of the History of Ideas, 183. Dordrecht and Boston: Kluwer Academic Publishers.

MANUEL, Frank E. 1974. *The religion of Isaac Newton*. Cambridge: Cambridge University Press.

MANDELBROTE, Scott. 1993. A Duty of the Greatest Moment: Isaac Newton and the Writing of Biblical Criticism. *The British Journal for the History of Science* 26, 3: 281-302.

———. 2007. Isaac Newton and the exegesis of the Book of Daniel. Pp. 351-375 in *Die Geschichte der Daniel-Auslegung in Judentum, Christentum und Islam*, K. Bracht and D.S. du Toit, eds. Berlin: Walter de Gruyter.

NEWTON, Isaac. undated(a). Yahuda Ms 1.1 (unpublished manuscript). Jerusalem: Jewish National and University Library.

———. undated(b). Irenicum (unpublished manuscript). Cambridge: King's College.

———. undated(c). Notes from Ramon Lull (unpublished manuscript). Stanford: California, Stanford University Library.

———. undated(d). Experimenta Raymundi (unpublished manuscript). Cambridge: King's College.

———. undated(e). Tabula Smaragdina and Hieroglyphica Planetarum (unpublished manuscript). Cambridge: King's College.

———. ca.1680s(a). Introduction to the Lexicon of the Prophets, Part two: About the appearance of the Jewish Temple (unpublished manuscript). Babson Ms 0434. Wellesley, MA: Babson College.

———. ca. 1680s(b). Drafts Concerning Solomon's Temple and the Sacred Cubit (unpublished manuscript). Yahuda Ms 2.4. Jerusalem: Jewish National and University Library.

———. ca. mid-1680s. The First Book Concerning the Language of the Prophets (unpublished manuscript). Keynes Ms 5. Cambridge: King's College.

———. 1690s. The Origins of Religions (unpublished manuscript). Yahuda Ms 41. Jerusalem: Jewish National and University Library.

————. 1737. *A Dissertation upon the Sacred Cubit of the Jews*. In *Miscellaneous Works of John Greaves Professor of Geometry at Oxford*. London.

————. 1957. Of an universall language. Translation and commentary by Ralph W. V. Elliott. *The Modern Language Review* III, 1 (1957): 1-18.

————. 1988. *The Chronology of Ancient Kingdoms Amended*. London: Histories & Mysteries of Man.

POPKIN, Richard Henry. 1992. *The Third Force in Seventeenth-century Thought*. Leiden and New York: E. J. Brill.

VILLALPANDO, Juan Bautista and Jerónimo del Prado. 1604. *Ezechielem Explanationes Et Apparatus Urbis Hierolymitani Commentariis Et Imaginibus Illustratus*. Rome.

VITRUVIUS. 1960. *The Ten Books on Architecture*. Morris Hicky Morgan, trans. New York: Dover Publications.

WESTFALL, Richard S. 1980. *Never at Rest: A Biography of Isaac Newton*. Cambridge: Cambridge University Press.

About the Author

Dr Tessa Morrison is an Australian Research Council post-doctoral Fellow in the School of Architecture and Build Environment at the University of Newcastle, Australia, Her academic background is in history, mathematics and philosophy and she has published articles on geometric and spatial symbolism, and architectural history. Her current research project is "Isaac Newton's Temple of Solomon and his analysis of sacred architecture: An interpretation and discussion of Babson Manuscript 0434". This project focuses on Newton's architectural interests and places it into perspective with his other works and into the context of his times.

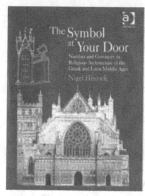

Book Review

Nigel Hiscock

The Symbol at Your Door: Number and Geometry in Religious Architecture of the Greek and Latin Middle Ages

Aldershot: Ashgate, 2007

Reviewed by Areli Marina

Keywords: church, baptistery, basilica, centrally-planned building, chapter house, Middle Ages, meaning, symbolism, geometry, harmonic proportion, measurement, number, Platonic solids, sacred geometry, shapes

School of Architecture
University of Illinois
117 Temple Buell Hall, MC 621
611 Lorado Taft Drive
Champaign, IL 61820 USA
amarina@illinois.edu

In his previous book, *The Wise Master Builder: Platonic Geometry in Plans of Medieval Abbeys and Cathedrals* (Aldershot: Ashgate, 2000), Nigel Hiscock sought to demonstrate that Platonic geometry underlay the ground plans of Western European Romanesque cathedrals and abbey churches, but refrained from arguing that these geometric and numerological foundations had symbolic meanings. In *The Symbol at Your Door*, he sets out to prove that medieval church design is pervaded with numbers and geometric forms that express metaphysical concepts central to Christian neo Platonism. In Hiscock's view, this symbolism was deliberately inscribed into the structures by the buildings' designers at the behest of their patrons. He argues that religious buildings that can be parsed into measures that correspond to Pythagorean number theory and broken down into fragments that coincide with Platonic geometry exist in such large numbers that the correspondences cannot be coincidental, but rather must have been intended by either patrons or builders. Furthermore, because there is so much evidence that well-educated medieval persons associated particular numbers, ratios, and geometric forms with Christian significance (such as the association of the number three with the Trinity, the square with the shape of the Holy Jerusalem, or the sphere with heaven), Hiscock concludes that patrons and designers must have meant to elicit those meanings in the mind of their viewers.

As the author acknowledges, this controversial hypothesis contradicts an influential strain of medieval architectural historiography that asserts that medieval builders used geometry and number purely for functional and not for representational purposes. This school of thought is exemplified by the work of Lon R. Shelby, John Harvey, François Bucher, or (more recently) Robert Bork, and emerges from the French nineteenth-century rationalist tradition of Eugene Emmanuel Viollet-le-Duc. Hiscock positions himself in a different genealogy, identifying his conceptual roots in the iconographic investigations of Richard Krautheimer, Erwin Panofsky, and Paul Frankl, among others. In fact, *The Symbol at Your Door* can be read as a prolonged meditation on questions suggested by Frankl in "The Secrets of the Mediaeval Masons" in 1945 (*Art Bulletin* 27,

n. 1, pp. 46-60): What means did medieval builders use to develop their buildings' plans, elevations, and decoration? How were these ideas communicated in the workshop? Why were the equilateral triangle and the square so important to medieval design practice? How were the metaphysical concepts about shape and measure presented by Plato in the *Timaeus* transmitted in the Middle Ages? What do these practices reveal about the relationship between the underlying geometry, workshop practice, aesthetic pleasure, and religious ideology? Although these are not new questions, most have never been conclusively resolved, and Hiscock's in-depth exploration of them is a welcome addition to the discourse.

The author divides his investigation into four parts. In the substantial Prologue, he examines the historiography of medieval architectural history and introduces the reader to Pythagorean and Platonic thought, its transmission in the Middle Ages, and its impact on architectural culture. This section digests material treated more fully in *The Wise Master Builder*. It outlines the relationship between arithmetic and the numbers perceived to be at the root of universal order (particularly 1, 2, 3, and 4) and the meanings attributed to them by ancient philosophers and Christian theologians. Chapters 1 and 2 constitute the second part of the book, in which the author explores the significance of two solid geometric forms, the sphere and the cube, as well as their deployment in cruciform churches. Chapters 3 through 6 are dedicated to the architectural use and meaning of plane figures: the equilateral triangle and the numbers derived from it (3, 6, 12); the square, square schematism, quadrature, and its associated forms, including the 1:√2 rectangle; the pentagon; and, in the last and longest chapter, circles, octagons, and other polygons. The book closes with an Epilogue in which the author summarizes his findings and considers their post-medieval implications.

To counter the arguments of the "rationalist school" and prove his hypothesis, Hiscock surveys and presents two bodies of classical and medieval textual sources and juxtaposes them with his analyses of medieval buildings. The first body of textual sources, including citations from authors such as Augustine, Boethius, and Dionysus, the Pseudo-Aeropagite, demonstrates that number and geometric form conveyed specific metaphysical meaning to educated audiences. The second shows that when writing ekphrases and other eulogistic texts on buildings, or using buildings as metaphors, medieval authors such as Eusebius and William Durandus often adopted the same numerological and geometric *topoi*. The range of sources cited by Hiscock is impressive; the bibliography includes 157 citations of ancient and medieval works. Relatively few of these writers, however, present direct evidence about a specific building at or around the time of its construction, and Hiscock focuses on those examples only briefly (pp. 37-38). He is most persuasive when discussing the abundance of evidence relating to a single building structure, such as Hagia Sophia in Istanbul (Chapter 1), or when he introduces new questions, such as why the expressive content of the stone tracery of stained glass windows is seldom considered when discussing their iconographic programs (Chapter 6).

The book's impressive breadth of scope is also its principal weakness. Hiscock conjures so many examples, cites so many authors, and illustrates so many buildings to prove his point that the vigor of his argument is dissipated in seemingly endless, often redundant, and occasionally contradictory exposition. Even the author seems to get lost in this thicket of evidence, suggesting, for example, that the burial and commemorative function of English cathedral chapter houses did not account for the choice of a polygonal plan on page 255, only to assert on page 257 that because chapter houses functioned as memorial shrines, they often imitated the octagonal form of baptisteries, mausolea, and martyria. Also, his persistent use of the passive voice obscures the agency with which he wants to endow patrons, designers, and builders, and often obfuscates his meaning. Judicious editorial pruning would have focused the narrative, and resulted in a more legible and ultimately more convincing text.

In the end, Hiscock cannot quite dispel the lingering doubt that, although highly-educated clerics and patrons may have been well-versed in the metaphysical aspects of number and geometric form, there is little consistent proof that either designers, builders, or audiences were equally familiar with them, much less across several centuries and from the British Isles to Asia Minor. Indeed, in cases where texts directly addressing geometry and number in relation to active workshops survive, such as the well-known debate regarding the design of Milan cathedral, they are silent on the subject of meaning and symbolic content (see especially pp. 366-370).

The philosopher Karl Popper has suggested that scientific theories cannot be proven true; at best, scientists can only refute untrue hypotheses. The volume of evidence amassed by Hiscock powerfully suggests, but does not prove, a strong causal relationship between the existence of Pythagorean and Platonic numbers and figures in medieval religious buildings and the symbolic interpretations of the buildings and building components made by their contemporaries. As in Frankl's day, the weak link remains the connection between the mind of the patron and the builder's hand (Frankl, 50). Insights into this aspect of the problem may lie, not within the purview of architectural history, but in the hands of our colleagues exploring the histories of medieval education and science.

In sum, *The Symbol at Your Door* is an ambitious book that is not afraid to engage a complex and controversial question. Hiscock's introduction to the metaphysics of number in the medieval period and the veritable anthology of medieval writings on architecture that he has assembled will be useful launching pads for future studies of how (or perhaps whether) the numbers and geometric forms embedded in specific medieval religious buildings had symbolic meaning. The wide range of his analyses reminds us of the richness and diversity of signification that architectural forms convey to their past and present audiences.

About the reviewer

Areli Marina, who trained as an art historian at the Institute of Fine Arts, New York University, teaches medieval and Renaissance architectural history at the University of Illinois. Her research focuses on the intersection of public rhetoric, national identity, and civic art production, particularly in relation to the semiotics of architecture and urban form; the problematic historiography of the Romanesque, Gothic, and Renaissance styles; and the role of antiquity in medieval and Renaissance art and architecture, with particular emphasis on northern Italy. Her work has been supported by a Rome Prize Fellowship at the American Academy in Rome, a Getty Foundation Fellowship, the Gladys Krieble Delmas Foundation, and the University of Illinois. Dr. Marina has recently completed a book on the medieval piazza, The Italian Piazza Transformed: Parma's City Center in the Communal Age, which will be published by Pennsylvania State University Press. During 2010-11, she hopes to make significant progress on a new book on medieval and Renaissance baptisteries while a Villa I Tatti fellow at the Harvard Center for Italian Renaissance Studies in Florence.

Conference Report

Architecture and Mathematics.

A seminar to celebrate Professor emeritus Staale Sinding-Larsen's 80th birthday

Trondheim, Norway, 25 November 2009

Report by Eir Grytli

Faculty of Architecture and Fine Art
Norwegian University of Science and Technology
Trondheim, NORWAY
eir.grytli@ntnu.no

Keywords: architecture and mathematics, Staale Sinding-Larsen

Trained as an art historian, Professor Staale Sinding-Larsen held the chair of architectural history at the Norwegian University of Science and Technology (NTNU, former NTH), from 1970 to 2000. During this period he also served for seven years as Director General of the Norwegian Institute in Rome. Despite his age, Professor Sinding-Larsen is still an active researcher and writer, with a continuing curiosity and open-mindedness towards the advantages of cross-disciplinary approaches to architectural research. On the occasion of his 80th anniversary, his colleagues at the Faculty of Architecture and Fine Art wished to honour him with a seminar focusing on one of the topics that is closest to Professor Sinding-Larsen's heart, namely the relationship between architecture and mathematics. In order to make the interdisciplinary research on the meeting-point between architecture and mathematics visible, the seminar was organized in co-operation between the Faculty of Architecture and Fine Art and the Department of Mathematical Sciences at NTNU.

Professor **Knut Einar Larsen**, the initiator and organizer of the seminar, opened the event. He emphasized the open-mindedness with which Staale Sinding-Larsen has always approached his own profession, art history, and how his curiosity had introduced new ways of understanding and assessing architecture, ways which still today influence the way architectural history is taught at NTNU.

The programme of the seminar was organized in two sections. The first section, *Historical Reflections*, presented research about how use of mathematical thought and methods (and particularly geometry) has influenced architectural design in the past. The second section, *The Contemporary Scene*, focused on how mathematics is used in contemporary architectural design, especially regarding the possibilities made available by computer design.

Sylvie Duvernoy opened the first session, speaking on the topic *Roman Architecture and Greek Mathematics – a Case Study: Pompeii's Amphitheatre*.

Sylvie Duvernoy trained as an architect in Paris, and earned a doctorate from the University of Florence in 1998. She is professor of architectural drawing at the University of Ferrara.

Nexus Network Journal 12 (2010) 357–360   NEXUS NETWORK JOURNAL – VOL.12, No. 2, 2010 **357**
DOI 10.1007/s00004-010-0036-2; *published online* 26 June 2010
© 2010 Kim Williams Books, Turin

Participants at the seminar to celebrate Professor emeritus Staale Sinding-Larsen's 80th birthday: 1-Staale Sinding-Larsen; 2-Knut Einar Larsen; 3-Tore Haugen; 4-Finn Hakonsen; 5-Sylvie Duvernoy ; 6- Fabian Scheurer; 7-Georg Glaeser; 8-Dag Nilsen; 9-Sverre Smalø. (photos by Georg Glaeser)

Duvernoy presented a case study on the amphitheatre in Pompeii. The study is a part of her comprehensive research of Roman amphitheatres, aimed at revealing the use of geometric diagrams in their designs through measured surveys and mathematical analysis of their curves. Around the beginning of the second century B.C the amphitheatre appeared as an architectural novelty in Roman culture, and was characterized by a closed elliptic shape that had never been previously adopted in architectural design. The period of time corresponds to the culmination of the Golden Age of Greek mathematics. It may be hypothesised that the necessity of designing a new building type provided theoretical mathematics with a successful field of direct and immediate experimentation. In her presentation, Duvernoy also called attention to how beauty is a related concept in architecture and in mathematics/geometry.

Dag Nilsen presented research entitled *A Mathematical Game. Studying ratios of measures as a possible method in building archaeology?* Dag Nilsen is trained as an architect, educated at NTNU (former NTH). Nilsen is associate professor at the Faculty of Architecture and Fine Art, NTNU, where he teaches architectural history and conservation.

When trying to reveal the history of a building, written sources are rarely sufficient to determine either its form and outline in earlier stages, or the intention of the builders. Nilsen maintains that through examining the building itself, it may be possible to develop a method to interpret specific questions encountered in building archaeology. By analyzing ratios of measures in two Norwegian medieval churches, it appears to be possible to explain previously unsolved problems of their design. In extensive studies of vernacular Norwegian architecture, Nilsen has also found indications that mathematical/geometrical tools have been used for design of building types not normally associated with professional designers.

Finn Hakonsen gave a talk on the subject *Ornament and Geometry – The case of Tuse village church in Denmark.* Finn Hakonsen is an architect trained at the Royal Academy of Fine Art in Copenhagen. His teaching and his writing are mainly based on the theories of the Tectonic Culture in Architecture.

The Tuse church at northern Seeland was built around the year 1200, in the Romanesque period of the country, and is one of more than a thousand medieval stone village churches in Denmark. The study presented is based on theories of architect Mogens Koch and its point of departure was the typical consecration crosses (*hjulkors*) found in the church, which appear as circular ornaments, usually associated with religious symbolism. The presentation launched the theory that these ornaments also embody a proportional system of the building. The study reveals correlations between the ornament and the proportions of the built structure. This leads to the question of whether the ornament, in addition to carrying religious meaning, also is a carrier of information for the aesthetical and structural design, a recipe for the entire building process.

Georg Glaeser opened the afternoon session with a talk entitled *The Infinite Variety of Curved Surfaces.* Georg Glaeser is a professor in mathematics and geometry at the University of Applied Arts in Vienna. He is also the author of several books about computer-based geometry. In the lecture, he presented the geometric programming system "Open Geometry". With the help of a C++ compiler, the user is able to create images and animations with arbitrary geometric context. The system provides a large library that makes it possible to carry out virtually any geometric task. It provides a pedagogical approach to understanding how curved surfaces are described as a function of mathematical formulas. His lecture was a fascinating demonstration of the possibilities of simulating curved surfaces using computer-aided programming tools, as well as illustrating how mathematics can also have an aesthetical dimension. The demonstration did not focus specifically on architectural structures, but showed the possibilities of modelling curved forms by mathematical formulas in a digital programming system which can be utilized in computer-aided architectural design.

Fabian Scheurer presented *The realization of the impossible: Geometrical Challenges in Contemporary architectural Concepts.* Fabian Scheurer is a computer scientist and partner of the company Designtoproduction in Zürich.

Today, mathematical and geometrical knowledge is of fundamental importance for the design and realization of complex, curved forms which are increasingly used in contemporary architecture design. Computer-aided technology has expanded the possibilities for architectural design and especially through the technology providing direct transfer from the design phase to production of components and elements, offering solutions for irregular shapes that otherwise could never have been realized. In his lecture, Scheurer also called attention to the challenges comprised between the seemingly infinite design possibilities and the complexity of the production and construction process, and how important it is that the computer tool be understood by the architects using it.

Finally, the central person of the day, **Staale Sinding-Larsen**, shared his *Meta-perspective on architecture and mathematics,* based on his view of an object; he dealt not with an object in itself, but rather what can be done with it. This approach to understanding objects can be referred back to the object-oriented paradigm in the digital world: an object consists of a set of attributes and methods. Methods are groups of instructions with reference to the attributes. A model for understanding mathematical influence on architectural design based on this theory will be presented in his forthcoming publication, *Borromini's Spire.*

Due to the busy end-of term period, only a limited number of students of architecture were able to dedicate the whole day to the seminar. This was a pity, because knowledge about how mathematics has influenced architecture as a design tool – and continues to do so increasingly – is vital for both a historical understanding of the built environment and for future design processes. In his lecture, Fabian Scheurer emphasized the need for architects to understand digital tools in order to obtain optimal benefits in complex design tasks; such an understanding requires a deeper knowledge of mathematics than is normally taught in architecture schools.

Personally, not being an expert myself on mathematics in architecture, I found the seminar most interesting and relevant. The deeper understanding of how mathematics and geometry has been a creative force for the design of buildings through all times provides a language for "reading" historic built structures which adds to the historical knowledge about the creation of buildings. Maybe it also provides a better understanding of why contemporary changes and additions to historic buildings often look alien to the original, if the architect who planned the addition did not know the mathematical language used in the original building.

About the reviewer

Eir Grytli is an architect, with a Ph.D in vernacular development history (1993). She is Professor at the Faculty of Architecture and Fine Art, Norwegian University of Science and Technology (NTNU), Trondheim, Norway, where she teaches architectural history and building conservation.

Tomás García-Salgado

National Autonomous University of Mexico
Palacio de Versalles 200
Col. Lomas Reforma MÉXICO
D. F. C.P. 11930
tgsalgado@perspectivegeometry.com

Erratum

Erratum to:

The Sunlight Effect of the Kukulcán Pyramid or The History of a Line

Erratum to: Nexus Network Journal 12 (2010) 113-130

DOI 10.1007/s00004-010-0019-3

The present erratum corrects an error in fig. 12, p. 124 of "The Sunlight Effect of the Kukulcán Pyramid or The History of a Line".

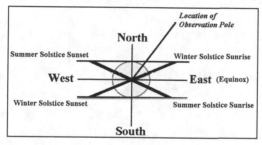

Original fig. 12

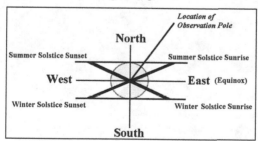

Corrected fig. 12 (Summer and Winter Soltice sunrise positions inverted)

The online version of the original article can be found under doi:10.1007/s00004-010-0019-3.

DOI 10.1007/s00004-010-0037-1; *published online* 6 July 2010
© 2010 Kim Williams Books, Turin

NEXUS NETWORK JOURNAL Architecture and Mathematics

Subscription information

ISSN print edition 1590-5896
ISSN electronic edition 1522-4600

Subscription rates

For information on subscription rates please contact:
Springer Customer Service Center GmbH
The Americas (North, South, Central America and the Caribbean)
journals-ny@springer.com
Outside the Americas: subscriptions@springer.com

Orders and inquiries

The Americas (North, South, Central America and the Caribbean)
Springer Journal Fulfillment
P.O. Box 2485, Secaucus, NJ 07096-2485, USA
Tel.: 800-SPRINGER (777-4643), Tel.:+1-201-348-4033
(outside US and Canada), Fax:+1-201-348-4505
e-mail: journals-ny@springer.com

Outside the Americas

via a bookseller or
Springer Customer Service Center GmbH
Haberstrasse 7, 69126 Heidelberg, Germany
Tel.: +49-6221-345-4304, Fax: +49-6221-345-4229
e-mail: subscriptions@springer.com
Business hours: Monday to Friday
8 a.m. to 8 p.m. local time and on German public holidays

Cancellations must be received by September 30 to take effect at the end of the same year.

Changes of address: Allow six weeks for all changes to become effective. All communications should include both old and new addresses (with postal codes) and should be accompanied by a mailing label from a recent issue.

According to § 4 Sect. 3 of the German Postal Services Data Protection Regulations, if a subscriber's address changes, the German Post Office can inform the publisher of the new address even if the subscriber has not submitted a formal application for mail to be forwarded. Subscribers not in agreement with this procedure may send a written complaint to Customer Service Journals, within 14 days of publication of this issue.

Back volumes: Prices are available on request.

Microform editions are available from: ProQuest. Further information available at: http://www.il.proquest.com/umi/

Electronic edition

An electronic edition of this journal is available at springerlink.com

Advertising

Ms Raina Chandler
Springer, Tiergartenstraße 17
69121 Heidelberg, Germany
Tel.:+49-62 21-4 87 8443
Fax:+49-62 21-4 87 68443
springer.com/wikom
e mail: raina.chandler@springer.com

Instructions for authors

Instructions for authors can now be found on the journal's website: birkhauser-science.com/NNJ

Production

Springer, Petra Meyer-vom Hagen
Journal Production, Postfach 105280,
69042 Heidelberg, Germany
Fax: +49-6221-487 68239
e-mail: petra.meyervomhagen@springer.com
Typesetter: Scientific Publishing Services (Pvt.) Limited, Chennai, India
Printers: Krips, Meppel, The Netherlands
Printed on acid-free paper
Springer is a part of
Springer Science+Business Media
springer.com
Ownership and Copyright
© Kim Williams Books 2010